Silk

The Textile Institute and Woodhead Publishing

The Textile Institute is a unique organisation in textiles, clothing and footwear. Incorporated in England by a Royal Charter granted in 1925, the Institute has individual and corporate members in over 90 countries. The aim of the Institute is to facilitate learning, recognise achievement, reward excellence and disseminate information within the global textiles, clothing and footwear industries.

Historically, The Textile Institute has published books of interest to its members and the textile industry. To maintain this policy, the Institute has entered into partnership with Woodhead Publishing Limited to ensure that Institute members and the textile industry continue to have access to high-calibre titles on textile science and technology.

Most Woodhead titles on textiles are now published in collaboration with The Textile Institute. Through this arrangement, the Institute provides an Editorial Board which advises Woodhead on appropriate titles for future publication and suggests possible editors and authors for these books. Each book published under this arrangement carries the Institute's logo.

Woodhead books published in collaboration with The Textile Institute are offered to Textile Institute members at a substantial discount. These books, together with those published by The Textile Institute that are still in print, are offered on the Woodhead web site at: www.woodheadpublishing.com. Textile Institute books still in print are also available directly from the Institute's web site at: www.textileinstitutebooks.com.

A list of Woodhead books on textiles science and technology, most of which have been published in collaboration with The Textile Institute, can be found towards the end of the Contents pages.

The team responsible for publishing this book

Commissioning Editor: Kathryn Picking
Project Editor: Cathryn Freear
Editorial and Production Manager: Mary Campbell
Production Editor: Mandy Kingsmill
Project Manager: Newgen Knowledge Works Pvt Ltd
Copyeditor: Newgen Knowledge Works Pvt Ltd
Proofreader: Newgen Knowledge Works Pvt Ltd
Cover Designer: Terry Callanan

Woodhead Publishing Series in Textiles: Number 149

Silk

Processing, properties and applications

K. Murugesh Babu

Oxford Cambridge Philadelphia New Delhi

Published by Woodhead Publishing Limited in association with The Textile Institute
Woodhead Publishing Limited, 80 High Street, Sawston, Cambridge CB22 3HJ, UK
www.woodheadpublishing.com
www.woodheadpublishingonline.com

Woodhead Publishing, 1518 Walnut Street, Suite 1100, Philadelphia, PA 19102-3406, USA

Woodhead Publishing India Private Limited, 303, Vardaan House, 7/28 Ansari Road, Daryaganj, New Delhi - 110002, India
www.woodheadpublishingindia.com

First published 2013, Woodhead Publishing Limited

British Library Cataloguing in Publication Data
A catalogue record for this book is available from the British Library.

Library of Congress Control Number: 2013939412

ISBN 978-1-78242-155-9 (print)
ISBN 978-1-78242-158-0 (online)
ISSN 2042-0803 Woodhead Publishing Series in Textiles (print)
ISSN 2042-0811 Woodhead Publishing Series in Textiles (online)

The publisher's policy is to use permanent paper from mills that operate a sustainable forestry policy, and which has been manufactured from pulp which is processed using acid-free and elemental chlorine-free practices. Furthermore, the publisher ensures that the text paper and cover board used have met acceptable environmental accreditation standards.

Typeset by Newgen Knowledge Works Pvt Ltd, India
Printed by Lightning Source

Contents

Woodhead Publishing Series in Textiles

1 **Watson's textile design and colour Seventh edition**
Edited by Z. Grosicki
2 **Watson's advanced textile design**
Edited by Z. Grosicki
3 **Weaving Second edition**
P. R. Lord and M. H. Mohamed
4 **Handbook of textile fibres Volume 1: Natural fibres**
J. Gordon Cook
5 **Handbook of textile fibres Volume 2: Man-made fibres**
J. Gordon Cook
6 **Recycling textile and plastic waste**
Edited by A. R. Horrocks
7 **New fibers Second edition**
T. Hongu and G. O. Phillips
8 **Atlas of fibre fracture and damage to textiles Second edition**
J. W. S. Hearle, B. Lomas and W. D. Cooke
9 **Ecotextile '98**
Edited by A. R. Horrocks
10 **Physical testing of textiles**
B. P. Saville
11 **Geometric symmetry in patterns and tilings**
C. E. Horne
12 **Handbook of technical textiles**
Edited by A. R. Horrocks and S. C. Anand
13 **Textiles in automotive engineering**
W. Fung and J. M. Hardcastle
14 **Handbook of textile design**
J. Wilson
15 **High-performance fibres**
Edited by J. W. S. Hearle
16 **Knitting technology Third edition**
D. J. Spencer
17 **Medical textiles**
Edited by S. C. Anand
18 **Regenerated cellulose fibres**
Edited by C. Woodings
19 **Silk, mohair, cashmere and other luxury fibres**
Edited by R. R. Franck

20 **Smart fibres, fabrics and clothing**
Edited by X. M. Tao
21 **Yarn texturing technology**
J. W. S. Hearle, L. Hollick and D. K. Wilson
22 **Encyclopedia of textile finishing**
H-K. Rouette
23 **Coated and laminated textiles**
W. Fung
24 **Fancy yarns**
R. H. Gong and R. M. Wright
25 **Wool: Science and technology**
Edited by W. S. Simpson and G. Crawshaw
26 **Dictionary of textile finishing**
H-K. Rouette
27 **Environmental impact of textiles**
K. Slater
28 **Handbook of yarn production**
P. R. Lord
29 **Textile processing with enzymes**
Edited by A. Cavaco-Paulo and G. Gübitz
30 **The China and Hong Kong denim industry**
Y. Li, L. Yao and K. W. Yeung
31 **The World Trade Organization and international denim trading**
Y. Li, Y. Shen, L. Yao and E. Newton
32 **Chemical finishing of textiles**
W. D. Schindler and P. J. Hauser
33 **Clothing appearance and fit**
J. Fan, W. Yu and L. Hunter
34 **Handbook of fibre rope technology**
H. A. McKenna, J. W. S. Hearle and N. O'Hear
35 **Structure and mechanics of woven fabrics**
J. Hu
36 **Synthetic fibres: nylon, polyester, acrylic, polyolefin**
Edited by J. E. McIntyre
37 **Woollen and worsted woven fabric design**
E. G. Gilligan
38 **Analytical electrochemistry in textiles**
P. Westbroek, G. Priniotakis and P. Kiekens
39 **Bast and other plant fibres**
R. R. Franck
40 **Chemical testing of textiles**
Edited by Q. Fan
41 **Design and manufacture of textile composites**
Edited by A. C. Long
42 **Effect of mechanical and physical properties on fabric hand**
Edited by H. M. Behery
43 **New millennium fibers**
T. Hongu, M. Takigami and G. O. Phillips

44 **Textiles for protection**
Edited by R. A. Scott
45 **Textiles in sport**
Edited by R. Shishoo
46 **Wearable electronics and photonics**
Edited by X. M. Tao
47 **Biodegradable and sustainable fibres**
Edited by R. S. Blackburn
48 **Medical textiles and biomaterials for healthcare**
Edited by S. C. Anand, M. Miraftab, S. Rajendran and J. F. Kennedy
49 **Total colour management in textiles**
Edited by J. Xin
50 **Recycling in textiles**
Edited by Y. Wang
51 **Clothing biosensory engineering**
Y. Li and A. S. W. Wong
52 **Biomechanical engineering of textiles and clothing**
Edited by Y. Li and D. X-Q. Dai
53 **Digital printing of textiles**
Edited by H. Ujiie
54 **Intelligent textiles and clothing**
Edited by H. R. Mattila
55 **Innovation and technology of women's intimate apparel**
W. Yu, J. Fan, S. C. Harlock and S. P. Ng
56 **Thermal and moisture transport in fibrous materials**
Edited by N. Pan and P. Gibson
57 **Geosynthetics in civil engineering**
Edited by R. W. Sarsby
58 **Handbook of nonwovens**
Edited by S. Russell
59 **Cotton: Science and technology**
Edited by S. Gordon and Y-L. Hsieh
60 **Ecotextiles**
Edited by M. Miraftab and A. R. Horrocks
61 **Composite forming technologies**
Edited by A. C. Long
62 **Plasma technology for textiles**
Edited by R. Shishoo
63 **Smart textiles for medicine and healthcare**
Edited by L. Van Langenhove
64 **Sizing in clothing**
Edited by S. Ashdown
65 **Shape memory polymers and textiles**
J. Hu
66 **Environmental aspects of textile dyeing**
Edited by R. Christie
67 **Nanofibers and nanotechnology in textiles**
Edited by P. Brown and K. Stevens

68 **Physical properties of textile fibres Fourth edition**
W. E. Morton and J. W. S. Hearle
69 **Advances in apparel production**
Edited by C. Fairhurst
70 **Advances in fire retardant materials**
Edited by A. R. Horrocks and D. Price
71 **Polyesters and polyamides**
Edited by B. L. Deopura, R. Alagirusamy, M. Joshi and B. S. Gupta
72 **Advances in wool technology**
Edited by N. A. G. Johnson and I. Russell
73 **Military textiles**
Edited by E. Wilusz
74 **3D fibrous assemblies: Properties, applications and modelling of three-dimensional textile structures**
J. Hu
75 **Medical and healthcare textiles**
Edited by S. C. Anand, J. F. Kennedy, M. Miraftab and S. Rajendran
76 **Fabric testing**
Edited by J. Hu
77 **Biologically inspired textiles**
Edited by A. Abbott and M. Ellison
78 **Friction in textile materials**
Edited by B. S. Gupta
79 **Textile advances in the automotive industry**
Edited by R. Shishoo
80 **Structure and mechanics of textile fibre assemblies**
Edited by P. Schwartz
81 **Engineering textiles: Integrating the design and manufacture of textile products**
Edited by Y. E. El-Mogahzy
82 **Polyolefin fibres: Industrial and medical applications**
Edited by S. C. O. Ugbolue
83 **Smart clothes and wearable technology**
Edited by J. McCann and D. Bryson
84 **Identification of textile fibres**
Edited by M. Houck
85 **Advanced textiles for wound care**
Edited by S. Rajendran
86 **Fatigue failure of textile fibres**
Edited by M. Miraftab
87 **Advances in carpet technology**
Edited by K. Goswami
88 **Handbook of textile fibre structure Volume 1 and Volume 2**
Edited by S. J. Eichhorn, J. W. S. Hearle, M. Jaffe and T. Kikutani
89 **Advances in knitting technology**
Edited by K-F. Au
90 **Smart textile coatings and laminates**
Edited by W. C. Smith
91 **Handbook of tensile properties of textile and technical fibres**
Edited by A. R. Bunsell

92 **Interior textiles: Design and developments**
Edited by T. Rowe
93 **Textiles for cold weather apparel**
Edited by J. T. Williams
94 **Modelling and predicting textile behaviour**
Edited by X. Chen
95 **Textiles, polymers and composites for buildings**
Edited by G. Pohl
96 **Engineering apparel fabrics and garments**
J. Fan and L. Hunter
97 **Surface modification of textiles**
Edited by Q. Wei
98 **Sustainable textiles**
Edited by R. S. Blackburn
99 **Advances in yarn spinning technology**
Edited by C. A. Lawrence
100 **Handbook of medical textiles**
Edited by V. T. Bartels
101 **Technical textile yarns**
Edited by R. Alagirusamy and A. Das
102 **Applications of nonwovens in technical textiles**
Edited by R. A. Chapman
103 **Colour measurement: Principles, advances and industrial applications**
Edited by M. L. Gulrajani
104 **Fibrous and composite materials for civil engineering applications**
Edited by R. Fangueiro
105 **New product development in textiles: Innovation and production**
Edited by L. Horne
106 **Improving comfort in clothing**
Edited by G. Song
107 **Advances in textile biotechnology**
Edited by V. A. Nierstrasz and A. Cavaco-Paulo
108 **Textiles for hygiene and infection control**
Edited by B. McCarthy
109 **Nanofunctional textiles**
Edited by Y. Li
110 **Joining textiles: Principles and applications**
Edited by I. Jones and G. Stylios
111 **Soft computing in textile engineering**
Edited by A. Majumdar
112 **Textile design**
Edited by A. Briggs-Goode and K. Townsend
113 **Biotextiles as medical implants**
Edited by M. King and B. Gupta
114 **Textile thermal bioengineering**
Edited by Y. Li
115 **Woven textile structure**
B. K. Behera and P. K. Hari

116 **Handbook of textile and industrial dyeing. Volume 1: Principles, processes and types of dyes**
Edited by M. Clark

117 **Handbook of textile and industrial dyeing. Volume 2: Applications of dyes**
Edited by M. Clark

118 **Handbook of natural fibres. Volume 1: Types, properties and factors affecting breeding and cultivation**
Edited by R. Kozłowski

119 **Handbook of natural fibres. Volume 2: Processing and applications**
Edited by R. Kozłowski

120 **Functional textiles for improved performance, protection and health**
Edited by N. Pan and G. Sun

121 **Computer technology for textiles and apparel**
Edited by J. Hu

122 **Advances in military textiles and personal equipment**
Edited by E. Sparks

123 **Specialist yarn and fabric structures**
Edited by R. H. Gong

124 **Handbook of sustainable textile production**
M. I. Tobler-Rohr

125 **Woven textiles: Principles, developments and applications**
Edited by K. Gandhi

126 **Textiles and fashion: Materials design and technology**
Edited by R. Sinclair

127 **Industrial cutting of textile materials**
I. Vi umsone-Nemes

128 **Colour design: Theories and applications**
Edited by J. Best

129 **False twist textured yarns**
C. Atkinson

130 **Modelling, simulation and control of the dyeing process**
R. Shamey and X. Zhao

131 **Process control in textile manufacturing**
Edited by A. Majumdar, A. Das, R. Alagirusamy and V. K. Kothari

132 **Understanding and improving the durability of textiles**
Edited by P. A. Annis

133 **Smart textiles for protection**
Edited by R. A. Chapman

134 **Functional nanofibers and applications**
Edited by Q. Wei

135 **The global textile and clothing industry: Technological advances and future challenges**
Edited by R. Shishoo

136 **Simulation in textile technology: Theory and applications**
Edited by D. Veit

137 **Pattern cutting for clothing using CAD: How to use Lectra Modaris pattern cutting software**
M. Stott

138 **Advances in the dyeing and finishing of technical textiles**
M. L. Gulrajani

139 **Multidisciplinary know-how for smart textiles developers**
Edited by T. Kirstein
140 **Handbook of fire resistant textiles**
Edited by F. Selcen Kilinc
141 **Handbook of footwear design and manufacture**
Edited by A. Luximon
142 **Textile-led design for the active ageing population**
Edited by J. McCann and D. Bryson
143 **Optimizing decision making in the apparel supply chain using artificial intelligence (AI): From production to retail**
W. K. Wong, Z. X. Guo and S. Y. S. Leung
144 **Mechanisms of flat weaving technology**
V. Choogin, P. Bandara and E. Chepelyuk
145 **Innovative jacquard textile design using digital technologies**
F. Ng and J. Zhou
146 **Advances in shape memory polymers**
J. Hu
147 **Design of clothing manufacturing processes: A systematic approach to planning, scheduling and control**
J. Gersak
148 **Anthropometry, apparel sizing and design**
D. Gupta and N. Zakaria
149 **Silk: Processing, properties and applications**
K. Murugesh Babu
150 **Advances in filament spinning**
D. Zhang
151 **Designing apparel for consumers: The impact of body shape and size**
M. E. Faust and S. Carrier

1
Introduction to silk and sericulture

DOI: 10.1533/9781782421580.1

Abstract: This chapter reviews the different types of mulberry and non-mulberry species of silk moth. It discusses the cultivation of different varieties of mulberry (moriculture), the life cycle and rearing of silkworms as well as the harvesting of cocoons (sericulture). It also covers diseases of silkworms and physical characteristics of cocoons.

Key words: silkworm, mulberry and non-mulberry species, moriculture, sericulture.

1.1 Introduction

Silk is one of the oldest fibres known to man. It is an animal fibre produced by certain insects to build their cocoons and webs. Although many insects produce silk, only the filament produced by the mulberry silk moth *Bombyx mori* and a few others in the same genus is used by the commercial silk industry (Jolly *et al.*, 1979). The silk produced by other insects, mainly spiders, is used in a small number of other commercial applications, for example weapon and telescope cross-hairs and other optical instruments (Spring and Hudson, 2002).

Over the centuries, silk has been regarded as a highly valued textile fibre. Its qualities of strength, elasticity, softness, absorbency, affinity for dyes and adaptability to various forms of twisting continue to meet various applications. Because of its high (tensile) strength, lustre, durability and ability to bind chemical dyes, silk is still considered a leading textile material (Zarkoob *et al.*, 2000). Despite facing keen competition from man-made fibres, silk has maintained its supremacy in the production of luxury apparel and other high-quality goods (Robson, 1998). Silk fibres display unusual mechanical properties: they are strong, extensible and mechanically compressible (Matsumoto *et al.*, 2006). Silk is rightly called the 'queen of textiles' for its lustre and feel (Manohar Reddy, 2009). Silk's natural beauty and properties of comfort in warm weather and warmth during colder months have also made it ideal for high-fashion clothing. As a result there is significant research into the artificial production of silk fibres (Chen *et al.*, 2003).

Sericulture is an art of rearing silkworm for the production of cocoons which are the raw material for the production of silk. The major activities of sericulture comprises food-plant cultivation to feed the silkworms which spin silk cocoons and reeling the cocoons for unwinding the silk filament for value added benefits such as processing and weaving (Kumar *et al.*, 2001). Sericulture is ideally suited for improving the rural economy as it is practised as a subsidiary industry to agriculture. Recent research has also shown that sericulture can be developed as a highly rewarding agro-industry. Sericulture involves the cultivation of mulberry and production of cocoons to produce silk filaments. The best raw silk is obtained from the species of moth *B. mori*. Breeding of silkworm normally occurs once in a year but, under industrial conditions, eggs may be hatched three times a year. The female moth lays around 350–400 eggs, after which the moths die. As they are subject to hereditary infection, any eggs from infected moths are destroyed. Larvae of about 3 mm are hatched from the eggs. For about 20–30 days, they are carefully nurtured and are fed five times a day on chopped mulberry leaves. In the meantime, the larvae change their skin four times and are formed into caterpillars about 9 cm long. At this point they are ready to spin a cocoon, for which racks, clusters of twigs or straw are provided.

The caterpillars have small openings under their jaws called spinnerets, through which they secrete a protein-like substance. This substance solidifies when it comes in contact with air and the resulting filament is spun around the silkworm in a shape resembling the digit 8. The cocoon, which is about the size of a peanut shell, is completed. The filament is held together by sericin or silk gum. The life of the worm is ended by the process of 'stoving' or 'stifling' in which the cocoons are heated. Some of the cocoons are preserved so that the pupae or chrysalises inside them develops into moths for further breeding.

There are five major types of silk of commercial importance, obtained from different species of silkworms which in turn feed on a number of food plants. The main type is mulberry. Other varieties of silks are generally termed non-mulberry silks. India has the unique distinction of producing all these commercial varieties of silk.

1.2 Mulberry silk species

The bulk of the commercial silk produced in the world comes from this variety. Mulberry silk (Fig. 1.1) comes from the silkworm *B. mori L.,* which feeds solely on the leaves of the mulberry plant. These silkworms are completely domesticated and reared indoors. In India, the major mulberry-silk-producing states are Karnataka, Andhra Pradesh, West Bengal, Tamil Nadu and Jammu and Kashmir, which together account for 92% of country's total mulberry raw silk production.

1.1 Mulberry silk: (a) worm, (b) moth and (c) cocoons.

B. mori, the domesticated silkworm, has been reared for over 2000 years. During this long history many mutations have occurred. The process of mutation has resulted in a combination of various genes producing a large number of silkworm races. Silkworm races are classified on the basis of:

1. place of origin;
2. voltinism (the number of broods or generations produced in a year); and
3. moulting (the number of times the caterpillar sheds its skin before starting to spin silk).

1.2.1 Classification by place of origin

There are various races defined by place of origin:

- Indian;
- Japanese;
- Chinese and
- European.

Indian races

These races are indigenous to India and South East Asia. The larval stage is longer and they are resistant to high temperature and humidity. The size of the cocoon and larvae is small. In many cases, the cocoon is spindle shaped and the cocoon colour is green, yellow or white. The cocoon shell is thin. They are mainly multivoltines. The quality of the silk filament is good.

Thai silk is only one of the mulberry silkworm (*B. mori*) silks but it differs somewhat in appearance, and is yellower in colour. The filament is coarser and has more silk gum (e.g., up to 37%) than normal mulberry silk (e.g. 20–25%) (Dhavalikar, 1962; Zhou *et al.*, 2000). These characteristics cause Thai silk to have its own style after weaving. Thai silk products are mainly produced by domestic industries in the northern and north east part of Thailand.

Japanese races

These races are indigenous to Japan. The larvae are robust. The cocoon is peanut-shaped. The cocoon colour is usually white but some are also green or yellow. The ratio of double cocoons is higher. The quality of the silk filaments is inferior, being thick and short. They are univoltine or bivoltine.

Chinese races

These races are indigenous to China. The larvae are resistant to high temperature but not to high humidity. The larvae grow quickly. The cocoon shape is, in many cases, elliptical but sometimes spindle shaped. The cocoon colour is white, golden yellow, green, red or beige. The cocoon filament is fine and reelability is good. They are univoltine, bivoltine and multivoltine.

European races

These races are indigenous to Europe and Central Asia. Larvae are vulnerable to high temperature and high humidity. The cocoon size is typically big with a little constriction. Cocoon reelability is good.

1.2.2 Classification by voltinism

Voltinism is a biological term referring to the number of broods or generations an organism may produce in a year. Silkworms are classified into:

- univoltines (races producing one generation a year);
- bivoltines (races producing two generations a year) and
- multi or polyvoltines (races producing more than two generations a year).

Univoltines

These races have only one generation in a year. The larvae body size is large. The cocoon weight, shell weight, shell ratio and cocoon filament weight are high. The cocoon filament quality is good.

Bivoltines

These races have two generations in a year. The cocoon weight, cocoon shell weight, shell ratio and cocoon filament weight are less compared to univoltines. Larvae are more robust and more uniform compared to univoltines.

Multivoltines

These races have more than two generations in a year. The life cycle is short. Larvae are robust and can withstand high temperature. The cocoons size is small. The cocoon weight, shell ratio and cocoon filament weight are lower compared to bivoltines. Cocoon filament is fine.

1.2.3 Classification by moulting

Based on moulting characteristics, silkworms are classified into:

- trimoulters;
- tetra-moulters;
- pentamoulters and
- (more rarely) bimoulters and hexamoulters.

Tetramoulters are mainly reared for commercial purposes.

1.3 Non-mulberry silk species

A large number of species (400–500) are used in the production of non-mulberry silks, but only about 80 have been commercially exploited in Asia and Africa, chiefly in tribal communities. The major varieties of non-mulberry silk are described below (Jolly *et al.*, 1979).

1.3.1 Tasar

Tasar (tussah) is coarse silk of copperish colour, mainly used for furnishings and interiors. It is less lustrous than mulberry silk, but has its own feel and appeal. Tasar silk (Fig. 1.2) is generated by the silkworm *Antheraea mylitta*, which thrives on the food plants Asan and Arjun. Silkworms are reared on trees in the open. In India, tasar silk is mainly produced in the states of Jharkhand, Chattisgarh and Orissa, besides Maharashtra, West Bengal and Andhra Pradesh. Tasar is an important source of income for many tribal communities in India.

1.3.2 Oak tasar

Oak tasar is a finer variety of tasar. In India oak tasar is produced by the silkworm *Antheraea proyeli J.* (Fig. 1.3). This species feeds on oak, found in abundance in the sub-Himalayan belt of India, covering the states of Manipur, Himachal Pradesh, Uttar Pradesh, Assam, Meghalaya and Jammu

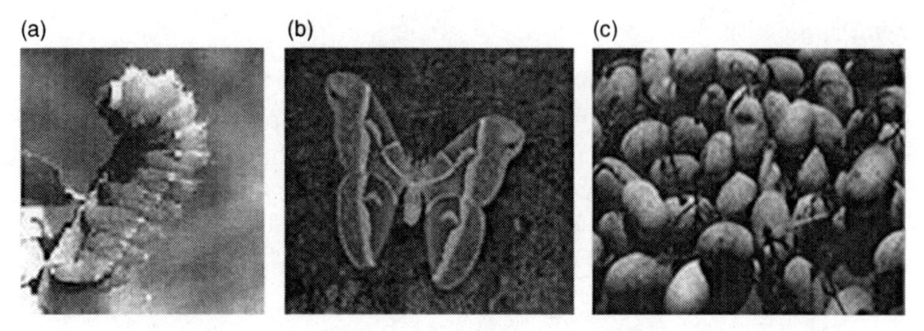

1.2 Tasar silk: (a) worm, (b) moth and (c) cocoons.

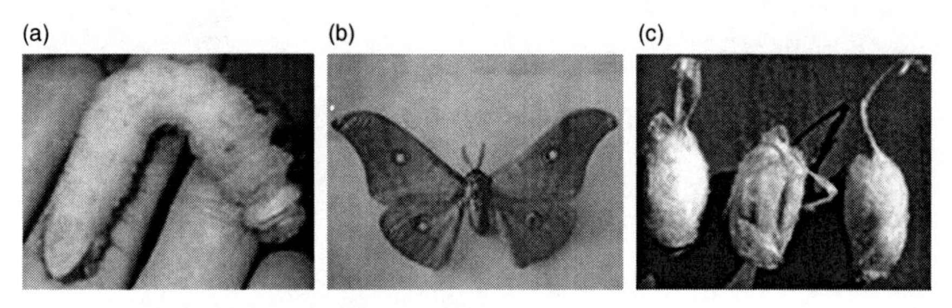

1.3 Oak tasar silk: (a) worm, (b) moth and (c) cocoons.

and Kashmir. China is the major international producer of oak tasar. Chinese oak tasar is produced by the silkworm *Antheraea pernyi.*

1.3.3 Eri

Eri is also known as Endi or Errandi, Eri is a multivoltine silk spun from open-ended cocoons, unlike other varieties of silk. Eri silk (Fig. 1.4) is the product of the domesticated silkworm, *Philosamia ricini* which feeds mainly on castor leaves. Like tasar, the cocoon varies in colour, size and softness. The soft cocoons are better for mechanical spinning. Harder and bigger cocoons are more appropriate for hand spinning (Sonwalker, 1969). In India eri is cultivated mainly in the north-eastern states and Assam. It is also found in Bihar, West Bengal and Orissa (Dhavalikar, 1962). Eri silk has both the softness of other silks and the insulating properties of wool, making it a promising silk species for further commercial development.

1.3.4 Muga

This golden yellow silk is the prerogative of India and is the pride of Assam state. It is obtained from semi-domesticated multivoltine silkworm,

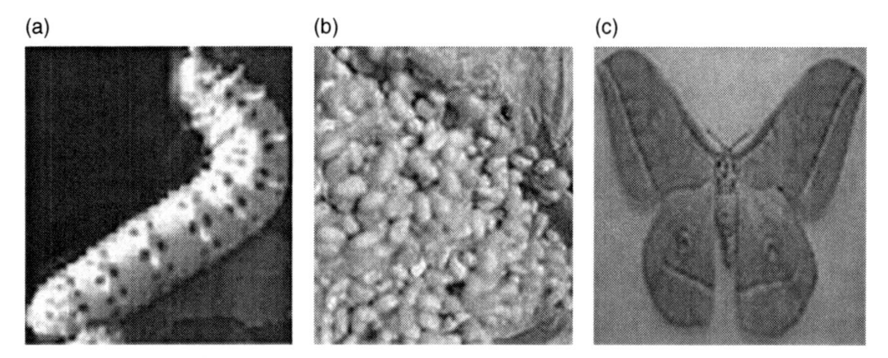

1.4 Eri silk: (a) worm, (b) cocoons and (c) moth.

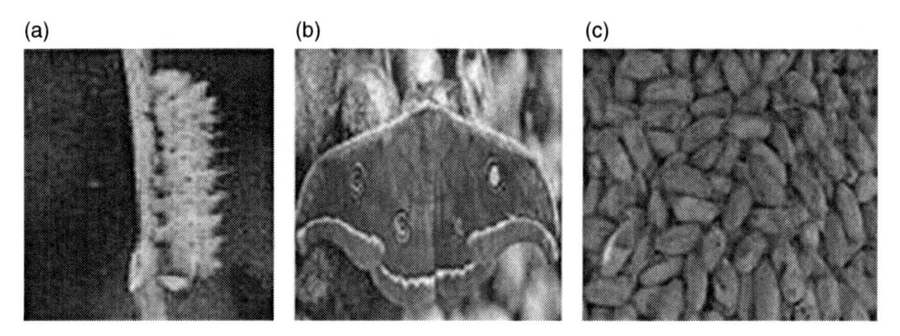

1.5 Muga silk: (a) worm, (b) moth and (c) cocoons.

Antheraea assamensis. These silkworms (Fig. 1.5) feed on the aromatic leaves of Som and Soalu plants and are reared on trees similar to that of the tasar. Muga culture is specific to the state of Assam and is an integral part of the tradition and culture of that state. Muga silk is a high-value product and is used in the manufacture of products such as sarees, mekhalas and chaddars.

1.3.5 Anaphe

This silk of southern and central Africa is produced by silkworms of the genus *Anaphe: A. moloneyi* Druce, *A. panda* (Boisduval), *A. reticulata* Walker, *A. carteri* Walsingham, *A. venta* Butler *and A. infracta* Walsingham. They spin cocoons in communes completely enclosed by a thin layer of silk. The tribal people collect them from the forest and spin the fluff into a raw silk that is soft and fairly lustrous. The silk obtained from *A. infracta* is known locally as 'book', and those from *A. moloneyi* as 'tissnian-tsamia' and 'koko'. The fabric is elastic and stronger than that of mulberry silk. Anaphe silk is used in velvet and plush.

1.3.6 Fagar

Fagar is obtained from the giant silk moth *Attacus atlas L.* and a few other related species or races inhabiting the Indo-Australian biographic region, China and the Sudan. They spin light-brown cocoons nearly 6 cm long with penduncles of varying lengths (2–10 cm).

1.3.7 Coan

The larvae of *Pachypasa otus D.*, from the Mediterranean region (southern Italy, Greece, Romania, Turkey, etc.), feed primarily on trees such as pine, ash, cypress, juniper and oak. They spin white cocoons measuring about 8.9 × 7.6 cm. In ancient times this silk was used to make the crimson-dyed apparel worn by the dignitaries of Rome; however, commercial production came to an end long ago because of the limited output and the emergence of superior varieties of silk.

1.3.8 Mussel

Where the non-mulberry silks previously described are of insect origin, mussel silk is obtained from a bivalve, *Pinna squamosa*, found in the shallow waters along the Italian and Dalmatian shores of the Adriatic. The strong brown filament or byssus is combed and then spun into a silk popularly known as 'fish wool'. Its production is largely confined to Taranto, Italy.

1.3.9 Spider silk

Spider silk is another non-insect variety. It is soft and fine, but also strong and elastic. The commercial production of this silk comes from certain Madgascan species, including *Nephila madagascarensis*, *Miranda aurentia* and *Eperia*. The spinning tubes (spinerules) are in the fourth and fifth abdominal parts to a frame, and the accumulated fibre is reeled out four or five times a month. Because of the high cost of production, spider silk is not used in the textile industry; however, durability, resistance to extremes of temperature and humidity make it indispensable for cross-hairs in optical instruments (Jolly *et al.*, 1979).

1.4 Types of mulberry and moriculture

Mulberry is a fast growing deciduous woody perennial plant. It has a deep root system. The leaves are simple, alternate, stipulate, petiolate, entire or lobed (Rangaswami *et al.*, 1976). The number of lobes varies from one to five. Plants are generally dioecious. Inflorescence is catkin with pendent or drooping peduncle bearing unisexual flowers. Inflorescence is always auxiliary. Male catkins are usually longer than the female catkins. Male flowers are loosely arranged and after shedding the pollen, the inflorescence dries and falls off. These are four persistent parianth lobes and four stamens implexed in bud. Female inflorescence is usually short and the flowers are very compactly arranged. There are four persistent parianth lobes. The ovary is one-celled and the stigma is bifid. The chief pollinating agent in mulberry is wind. Mulberry fruit is a sorosis, mainly violet black in colour. Most of the species of the genus *Morus* and cultivated varieties are diploid, with 28 chromosomes. However, triploids are also extensively cultivated for their adaptability, vigorous growth and quality of leaves.

Cultivation of mulberry plants is referred to as moriculture. Mulberry is a perennial plant and, once established in the field, continues to produce in full form for at least 15 years (Ullal and Narasimhanna, 1987). Though arboreal in nature, it is trained as low bush for commercial exploitation. Mulberry is a hardy plant and can tolerate a varied range of agro-climatic conditions. However, the most suitable range of conditions includes a temperature range of 20–30°C and sunshine of 9–13 h per day. On average, mulberry requires 50–60 mm of water per week. Under such conditions (which are the prevailing conditions in the southern tropics of India), five to six crops can be harvested in a year, whereas in temperate conditions, two to three crops are harvested.

1.4.1 Types of mulberry

There are about 68 species of the genus *Morus*. The majority of these occur in Asia, especially China (24 species) and Japan (19). Continental America is also rich in its *Morus* species (Sastry, 1984). The genus is poorly represented in Africa, Europe and the Near East, and it is not present at all in Australia. In India, there are many species of *Morus*, of which *M. alba, M. indica. M. serrata* and *M. laevigata* grow wild in the Himalayas. Several varieties have been introduced belonging to *M. multicaulis*, *M. nigra*, *M. sinensis* and *M. philippinensis*. Most of the Indian varieties of mulberry belong to *M. indica*. In China there are 15 species, of which 4 species, *Morus alba*, *M. multicaulis*, *M. atropurpurea* and *M. mizuho*, are cultivated for sericulture. In the former Soviet Union *M. multicaulis*, *M. alba, M. tartarica* and *M. nigra* are present.

1.4.2 Climatic requirements and soil conditions

The success of mulberry leaf production depends on three factors:

1. variety (including resistance to diseases and pests);
2. climatic conditions, which are essential for the growth of mulberry – different varieties require different climatic conditions and
3. cultivation practices (agronomic characters like good rooting, fast growth, high yield, soil conditions, plant protection measures).

Mulberry thrives under various climatic conditions ranging from temperate to tropical and is found north of the equator between 28°N and 55°N latitude. The ideal range of temperature is from 24°C to 28°C. Mulberry grows well in places with an annual rainfall ranging from 600 to 2500 mm. In areas with low rainfall, growth is limited through moisture stress, resulting in low yields. On average, mulberry requires 340 m^3/ha of water every 10 days in case of loamy soils and 15 days in clayey soils. Atmospheric humidity in the range of 65–80% is ideal for mulberry growth. Sunshine is one of the important factors controlling growth and leaf quality. In the tropics, mulberry grows with a sunshine range of 9–13 h a day. Mulberry can be cultivated from sea level up to an elevation of 1000 m.

Mulberry flourishes well in soils that are flat, deep, fertile, well drained, loamy to clayey and porous with good moisture holding capacity. The ideal range of soil pH is 6.2–6.8, the optimum being 6.5–6.8. Soil amendments may be used to obtain the required pH. Powdered gypsum/lime is mixed well with the soil and irrigated to stagnation for 48–72 h. Later the water is leached out by drainage and dried. The quantity of gypsum or lime applied in different cases to bring the pH to 6.8 are given in Table 1.1.

1.4.3 Mulberry planting methods

The two major methods of planting followed in countries like India are the Pit and the Row systems.

Pit system

This system is followed for rain-fed crops and adopts a wider spacing. Instead of ploughing the entire field, pits of standard size (40 × 40 × 40 cm) are dug with an inter-plant and inter-row distances of 90 × 90 cm for a bush type of cultivation, 180 × 90 cm for high bush cultivation and 270 × 270 cm for a tree-type plantation. Equal quantities of organic manure, red soil and sand are placed in each pit after mixing and a cutting or a sapling is planted. Initially it is watered daily until rooting takes place. If tree-type

Table 1.1 Gypsum or lime applications used to bring soil pH to 6.8

	Acid soils		Alkali soils	
	Lime applied (t/ha)			
pH range	Plain	Hilly areas	Soil type	Gypsum applied (t/ha)
5.5–6.5	1.25	2.5	Sandy	–
	2.50	5.0	Sandy loamy	–
	5.0	7.5	Loamy	–
	7.5	8.75	Clay loamy	–
7.4–7.8	–	–	–	2.0
7.9–8.4	–	–	–	5.0
8.5–9.0	–	–	–	9.0
≥9.1	–	–	–	14.0

1.6 Row system of mulberry planting.

plants are to be grown on hedges, roadside, etc., the pits are of a larger size (45 × 45 × 45 cm).

Row system

This system is followed for irrigated mulberry crops throughout South India. The land is prepared by ploughing ridges and furrows. The distance between the ridges is generally 45–60 cm. A rope with knots at equal distances of about 45–60 cm is tied from one end of a ridge to the opposite end, and two cuttings are planted at the point indicated by the knot on either side of the ridge. Thus, the inter-plant distance between the rows and plants within the row is about 45–60 cm (Fig. 1.6). Irrigation water flows through the furrows between the rows and generally the crop is grown as bush mulberry and is harvested by bottom pruning.

1.4.4 The main pests of mulberry

Leaf yield from mulberry becomes considerably reduced when the plant is attacked by diseases and pests. Further, leaves become nutritionally poor or sometimes entirely unfit for silkworm feeding. The main pests of mulberry are:

- *Maconellicoccus hirsutus* – mealy bug;
- *Diaphania pulverulentalis* – leaf roller and
- *Spilarctia obliqua* – bihar hairy caterpillar.

Minor pests are: thrips, jassids, scale insects and shorthorned grasshopper.

1.4.5 The main diseases of mulberry

Mulberry diseases may be infectious or non-infectious. Infectious diseases are caused by pathogens. Non-infectious diseases are those that are due to certain deficiencies in nutrients essential to the plant. Infectious diseases are classified into fungal, bacterial, viral and nematode from the causative organism. Of these, 10 or 12 diseases cause severe damage to mulberry plants. They affect different parts of the plant. The economy of sericulture is severely affected since both the quality and quantity of the leaf produced are affected by disease. To control the spread of the disease or pest over large areas and over a number of years, early diagnosis and immediate counter-measures are essential. Foliar diseases include leaf spot, leaf rust, powdery mildew, leaf blight and bacterial blight. Soil-borne diseases include root rot and root knot. Nursery diseases include stem canker, cutting rot, collar rot and dieback. Some of the most important are discussed below.

Foliar diseases: powdery mildew

Introduction: Powdery mildew is caused by the fungal pathogen *Phyllactinia corylea* and is more common in temperate regions. In the tropics, it is common during winter and the rainy season. Temperatures of 22–26°C and RH of 60–70% favour the spread of disease which is initially characterized by white powdery patches on the lower surface of the leaves. As the disease advances, the patches spread to the entire leaf surface and turn a blackish brown colour (Fig. 1.7).
Control: Improved aeration and sunlight in the mulberry garden will help to check the spread of the disease as will spraying 0.2% Carathian or 0.2% Bavistin fungicides on the lower surface of the leaf. In highly disease-prone areas it is advised to go for wider spacing and to plant powdery mildew tolerant mulberry varieties.

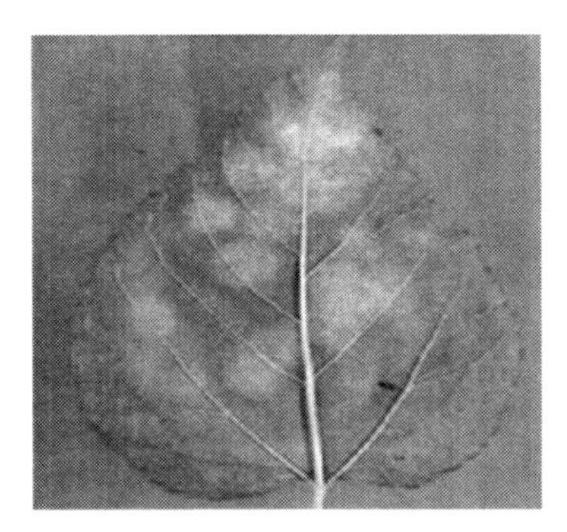

1.7 Mulberry diseases: powdery mildew.

Foliar diseases: leaf spot

Introduction: Leaf spot disease in mulberry is caused by the fungal pathogen *Cercospora moricola*. The incidence of disease increases during the rainy season. The diseased leaves have a number of circular or irregular brownish black spots of varying sizes. Severely affected leaves become yellowish and fall prematurely.
Control: To check the spread of the disease, it is advisable to go for plantation with a wider spacing. The infected leaves must be collected and burnt. Two sprays of 0.2% Bavistin at an interval of 10–15 days (safe period 4 days) will help to check the spread of the disease.

Soil diseases: root rot

Introduction: Root rot disease is increasingly prevalent and is caused by the fungus *Fusarium solani and F. oxysporum*. The disease appears in all types of soil and climate throughout the year. Initially, the disease appears in isolated patches and spreads quickly to surrounding areas. Infected plants show the symptoms of sudden withering of leaves followed by drying/death due to the decaying of the root (Fig. 1.8).
Control: Application of farmyard manure at 20 tonnes/hectare as basal and the pouring of copper oxy chloride (2 g/L of water) at the root surface prevents the spread of the disease to other healthy plants by basin irrigation. The dead plants should be uprooted and *Trichoderma viride* at 25 g/plant as well as *Bacillus subtilis* at 25 g/plant should be applied at the time of planting or pruning.

Soil diseases: root knot

Introduction: Root knot caused by a nematode, *Meloidogynae incognita*, is one of the most serious and widely observed diseases of mulberry. It is

1.8 Mulberry diseases: root rot.

more prevalent in sandy soil under an irrigated farming system. The disease is soil-borne in nature and spreads through contaminated saplings. The affected plants show stunted growth, marginal necrosis and chlorosis of leaves. The underground symptoms include the formation of knots/galls on the roots.

Control: Deep ploughing in summer, applying neem cake at 1000 kg/ha and Carbofuran 3G at 30 kg/ha/year in four split doses (safe period is 50 days).

1.5 The life cycle of the silkworm

The silkworm larval life is divided into five instars, separated by four moults (Fig. 1.9). Four distinct stages of development complete one generation of the species:

1. egg;
2. larvae;
3. pupa and
4. moth.

After hatching from the egg, larvae go through four moults as they grow. During each moult, the old skin is cast off and a new, larger one is produced.

1.5.1 Stage 1: egg (incubation 10–14 days)

The first stage of a silkworm's life cycle is the egg. The female moth lays an egg about the size of an ink dot during summer or early autumn. The egg remains

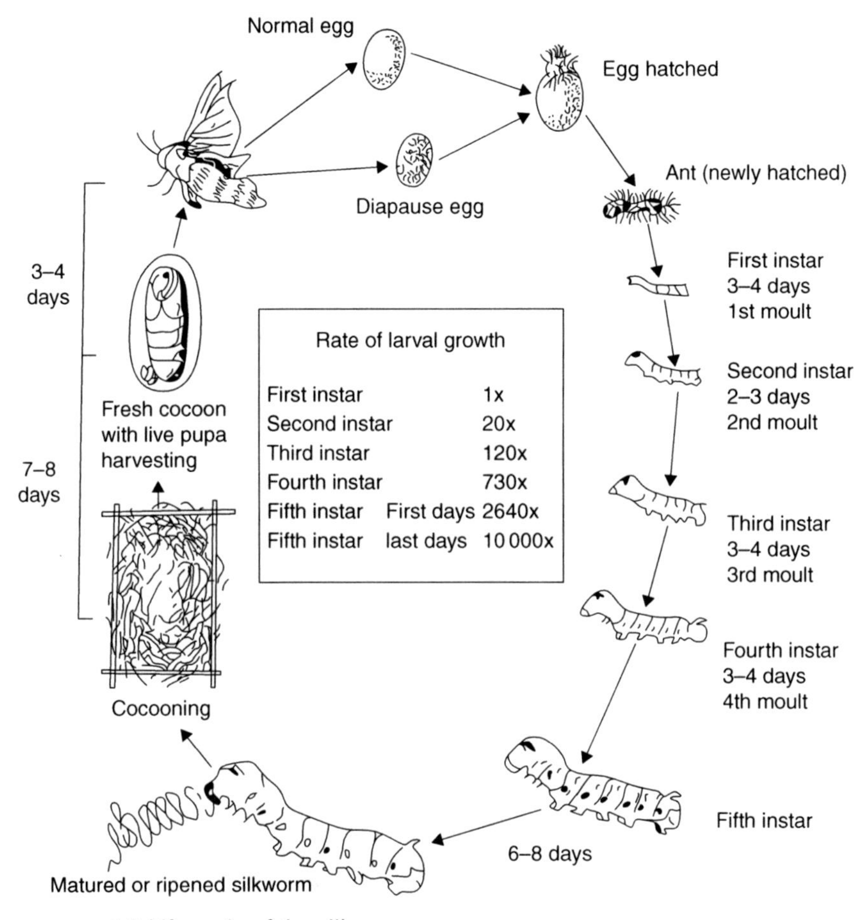

1.9 Life cycle of the silkworm.

in a dormant stage until spring arrives. The warmth of the spring stimulates the egg to hatch. Newly laid eggs are a creamy yellow, and after a few days the fertile live eggs will turn grey.

1.5.2 Stage 2: larva (27 days; five instars)

The silkworm, upon hatching, is about 1/8th of an inch long and is extremely hairy. Young silkworms can only feed on tender mulberry leaves. However, during the growth phase they can eat tougher mulberry leaves as well. The larval stage lasts for about 27 days and the silkworm goes through five growth stages called instars during this time. During the first moulting, the silkworm sheds all its hair and gains a smooth skin.

1.5.3 Stage 3: pupa (14 days)

Pupa or cocoon is the stage in which the larva spins silk threads around it in order to protect itself from its predators. The larva traps itself inside the cocoon in order to pupate. The colour of the cocoon varies, depending on what the silkworm has eaten. It can range from white to golden yellow. The second moulting occurs inside the cocoon, when the larva turns into a brown pupa. It takes about 2–3 weeks for the pupa to metamorphose into an adult moth. The silkworm will spin a silk cocoon as protection for the pupa. Cocoons are shades of white, cream and yellow. The glistening shine of the silk gives an impression of silver and shades of gold. After a final moult inside the cocoon, the larva changes into the brown pupa. Further changes inside the pupa result in an emerging moth.

1.5.4 Stage 4: moth (5–7 days)

An adult silk moth emerges from the cocoon about two weeks after completion. This is the adult stage of the silkworm *B. mori*. The body of the moth is covered in short fine hair and the wings are a creamy white colour with faint brown lines. Moths cannot fly or consume nutrition. Once the adult moth comes out of its cocoon, its only purpose is to find a member of the opposite sex and mate. Males are larger than females and more active. They flap their wings rapidly to attract females. Within 24 h of mating, the male moth dies, while the female lays abundant eggs, after which it dies as well.

Normally in nature, there is only one season per year during which silkworms reproduce. However, in countries like India and China silkworms reproduce continuously which is also the case when silkworms are bred for commercial purposes. They cannot eat or drink on their own so that everything from their food to their habitat has to be organized artificially. Although, silkworms feed on mulberry leaves, it is not always possible to find them, especially during autumn and winter. Commercial silkworm breeders therefore feed larvae a substance called silkworm chow, which is a good alternative to mulberry leaves.

1.5.5 Spinning of silk by silkworms

The spinning process and its mechanism have been studied by number of researchers. The natural silk synthesized by the silkworm and spun in the form of a silk cocoon is synthesized in the silk gland. The silk gland of *B. mori* is a typical exocrine gland secreting a large amount of silk proteins. It is a paired organ consisting of modified labial/salivary glands located at

1.10 Spinning of silk by silkworms.

the two lateral sides under the alimentary canal. Each gland is basically a tube made of glandular epithelium with two rows of cells surrounding the lumen (Mondal *et al*., 2007). In the gland of the silkworm *B. mori*, the protein fibroin exists in a concentrated solution in a random-coil conformation. The silk is secreted as a 15% aqueous solution of fibroin in the narrow, convoluted posterior gland. From there it moves to the middle storage gland where the fibroin is enveloped in a solution of sericin that will ultimately become the gummy coating which holds the fibres together in the cocoon. It is in the middle gland, where the concentration of silk rises to 30%, that the polymer is expected to be precipitated (Calvert, 1985). It is assumed that the molecules are not entangled, as in a solution of a typical synthetic polymer, but exist as separate particles within a solution of globular proteins. The silkworm extrudes the fluid through a spinneret of about 10 μm in diameter, and then stretches the fibre by pulling the spinneret away from a point where the thread has been attached to a support (Fig. 1.10). This second elongation (drawing) step results in a stiff fibre, with crystals orientated along its axis (Calvert, 1998).

The silk glands of the *B. mori* are structured like tubes consisting of posterior, middle and anterior sections. The posterior is long and thin. The middle is short with a diameter measuring 3–4 mm. The anterior is extremely thin, leading to the spinneret in the head of the larvae from which the silk is excreted. Fibroin is secreted in the posterior section and transferred by peristalsis to the middle section, which acts as a reservoir. Here it is stored as a viscous aqueous solution until required for spinning. The majority of the sericin is created within the walls of the middle section. In fact, these two proteins co-exist side by side in the middle section without mixing. The fibroin core is covered with a layer of sericin and the secretions from the two proteins join at the junctions where the sericin is fused into one layer. The Filipis glands discharge a liquid protein. To form its cocoon, the silkworm draws out the thread of liquid protein and internally adds layer after layer to complete this protective covering.

1.6 Early age rearing of silkworms

Rearing of young silkworms up to the second moult is called 'chawki rearing', which usually lasts up to 10 days and is a vital aspect of sericulture industry. This stage of larvae requires ideal environmental conditions and tender mulberry leaves. Robust growth and development of chawki larvae make them resistant to diseases and more stress tolerant during the later stages of development. Silkworm layings should be procured only from licensed seed preparers which are tested and certified as disease-free layings (dfls). Under normal recommended cultivation practices of mulberry 250–300 dfls can be brushed per crop per acre. Silkworm layings have to be transported in egg-carrying boxes, during the cooler hours of the day, from the grainages to the place of rearing. Transportation in egg transportation boxes avoids the exposure of layings to extremes of temperatures and humidity. During the incubation period, care should be taken to preserve the layings in a disease-free environment, with relative humidity of more than 80% and a temperature of 24–25°C. The embryo inside the eggs grows utilizing the reserved food materials in the yolk and hatches on the tenth day.

Chawki rearing may require technical skill which is not available among the common sericulturists. Further, many sericulturists do not have the necessary equipment to rear the young age silkworms under ideal conditions. These difficulties could be mitigated by rearing of young age worms in Chawki Rearing Centres (CRCs). In CRCs rearing of young age larvae is conducted by trained technicians with the requisite equipment for maintaining the optimum temperature and humidity. Mulberry leaf required for the young age worms is ensured. The worms are distributed to the farmers when they have settled for second moult or out of moult.

Chawki worms, reared with care under ideal conditions tend to grow vigorous and healthy and give stable crops when shifted to the rearers' house. This system cuts down the cost and also the rearing period at individual farmers' houses and is proven to have better returns. The concept of chawki rearing is not new. In developed countries like Japan, China and Korea, chawki worms are typically supplied to the farmers. For example, in Japan about 95% of the worms are distributed only after chawki rearing. On the contrary in India, only about 10% of the rearers receive chawki reared worms.

Chawki rearing can be done in different ways. The most popular methods are:

- **Stand rearing:** Chawki larvae are reared in plastic rearing trays of 3′ × 4′ size. The trays are arranged in rearing stands. Paraffin paper is used as the seat and cover.
- **Box rearing:** Wooden trays of 4 × 3 × 2 and 4 feet depth are used in this method. Trays during the feeding period are arranged one above the other to a convenient height, which can increase the temperature/humidity in the rearing bed. The trays are kept in a criss-cross arrangement for 30 min before feeding to allow fresh air to circulate. The paraffin or polythene sheets are removed 30 min before feeding and during the moulting period.
- **Plastic trays:** Plastic trays are also used for rearing. The trays are arranged one above the other on a chawki rearing stand. This method helps in maintaining the temperature and humidity.

Key aspects of rearing include the following:

- **Rearing house:** A separate rearing house with adequate rearing space, sufficient ventilation and light is essential. This will enable effective disinfection and the maintenance of humidity, temperature and other hygienic conditions. The rearing house should be away from dwellings.
- **Feeding:** Mulberry leaves should be harvested during cool hours of the day and preserved with wet gunny cloth to avoid leaf driage. Two to three feeds per day is recommended. Leaves need to be chopped to enable uniform feeding.
- **Spacing:** As the larvae grow, they require more bed area. Usually, young age rearing is conducted in crowded conditions.
- **Bed cleaning:** Leftover leaf and litter accumulated in the rearing bed increases the bed humidity and leads to multiplication of pathogens. Hence periodic bed cleaning is necessary. However, cleaning during the first instar is not recommended as it could increase the percentage of missing larvae. Cleaning is done twice during the second instar, once after resumption from the first moult and again before settling for the second moult. Cleaning nets should be used for bed cleaning to avoid secondary contamination.
- **Moulting care:** When the larvae start settling for moult, the top paraffin paper is removed, the bed is spread and feeding is stopped. A thin layer of lime dusting (at 3 g/sq ft) helps in drying of the beds, especially during the rainy season. When more than 95% of the larvae are out of moult, Vijetha, bed disinfectant (Vetcare) is dusted and after an hour feeding is resumed.

- **Transportation of chawki larvae:** Chawki worms are to be checked for diseases before distribution to farmers. Larvae are distributed to the farmers when they are settled for the second moult. The rearing bed is rolled and piled up in a tray for transportation, which should take place during the cooler hours of the day, preferably during evening.

1.7 Late age rearing of silkworms

The larval duration in the life cycle of the silkworm ranges from 24 to 28 days. The larval stage comprises five instars and four moults. Rearing of silkworms from the third or fourth instar up to spinning stage is called late age silkworm rearing. During this period, the silkworm consumes more than 94% of the total mulberry leaves required, increases in body size by a factor 133, increases in weight by a factor 125, and the silk gland weight increases by three orders of magnitude compared to the time of hatching. Since the late age silkworms are sensitive to high temperature and humidity, scientific rearing methods are necessary to achieve maximum growth and survival rate of the larvae, which in turn increases the cocoon yield and silk production.

1.7.1 Rearing house

A separate rearing house with good ventilation, light and space is constructed for late age rearing, with adequate provision for maintaining the required temperature and room humidity. The rearing room must be clean and hygienic and suitable for conducting effective disinfection. It should have an antechamber for the prevention of uzifly entry as well as a separate leaf storage room.

In tropical countries such as India, rearing houses should be constructed east–west longitudinally with corridors all around and windows on the south and north sides of the building. For easy exchange of air, the windows should be constructed opposite to one another. All the doors and windows should be fixed with uzifly mesh. Ventilators should be fixed above and below the windows. The rearing space required for 1 dfls is 8 sq ft. The rearing rooms and rearing appliances should be disinfected before and after the completion of silkworm rearing.

1.7.2 Silkworm rearing in bamboo trays

Rearing silkworms in bamboo rearing trays has been the traditional method of rearing in many countries (Fig. 1.11). Late age silkworm larvae require a temperature of 22–25°C, 70–80% RH and 16.08 h of light and darkness respectively. The rearing room must be adequately ventilated. Depending

1.11 Late age rearing in bamboo trays.

on the season and environment, the silkworms are to be fed three to four times per day. The mulberry leaves should be harvested during the cooler hours of the day and transported in bamboo leaf baskets covered with wet gunny cloth to the leaf storage room. The leaves are then stored in leaf storage bins which are covered with wet gunny cloth; this will preserve the leaf moisture for a longer period and maintain the quality of the leaves. Perforated PVC pipes can be placed inside the heap, which help to reduce temperature build-up inside the leaf chamber.

During moulting periods, the rearing bed has to be dried by dusting lime powder, and aeration inside the room should be increased by opening the windows and ventilators to the required extent. After more than 95% of the larvae come out of moult, rearing bed disinfection should be performed by dusting the bed disinfectants as per recommendation, and feeding is then resumed after 30 min. Any diseased larvae or unequal size larvae observed should be separated and disposed of in a basin containing formalin or bleaching powder solution. Regular rotation of the position of rearing trays from the top to the bottom will ensure uniform temperature and humidity to the larvae. Depending on the season and weather conditions, the ventilators and windows can be opened or closed to facilitate maintenance of the temperature and humidity inside the room.

1.7.3 Shoot feeding method of rearing

To bring down the labour cost and increase productivity, shoot feeding methods of rearing have been developed. These are standardized for tropical conditions and are recommended for the sericulturists. This new

1.12 Shoot rearing method.

technology reduces the labour cost by 50%. Irrigated mulberry gardens are suitable for adoption of shoot feeding method. Five to six crops can be harvested in a unit area of mulberry garden with an interval of 40–45 days between crops. In this method of rearing, a 15–20% leaf saving can be achieved which translates to an increase of cocoon productivity per unit area of land. Since the leaves are attached to the shoots, the quality of the leaves is maintained for a longer period which facilitates effective feeding and reduces the leaf cocoon ratio significantly. Mulberry shoots should be cut using sharp secateurs or a sharp sickle leaving two to three buds, and harvested during the cooler hours of the day. They should then be covered with wet gunny cloth and immediately transported to the leaf storage room. The shoots are placed upright and covered with wet gunny cloth.

In this method of rearing, bamboo trays or rearing stands are not required; instead shoot rearing racks are constructed (Fig. 1.12). The rack should be 5 ft in width, with a length as long as the rearing hall. The rearing rack can have three to four tiers with a gap of 2–2.5 ft between the racks for effective aeration. 600–750 sq ft rearing space is required for hybrids while the CSR hybrids require 800–850 sq ft bed area.

After resuming from the moult, the second or third instar larvae are to be evenly spread on lime-powder-dusted newspaper spread on the shoot rearing rack. Mulberry shoots should be gently placed on the rearing bed opposite to one another. Depending on the season, three to four feedings can be given per day and as the larvae grow, bed spacing should be increased. When more than 90% of the larvae settle for the fourth moult, feeding should be stopped, aeration inside the rearing room increased and rearing beds should be dried by dusting with active lime powder. When more than 90% of the larvae come out of the moult, the rearing bed should be disinfected by dusting with the recommended quality of disinfectants. After 30 min feeding should be resumed. In the shoot feeding method, the rearing bed is cleaned once after the fourth moult by using rearing nets or rope, larvae should then be spread evenly as per the recommendation, and feeding resumed. If the rearing room humidity is very low, the rearing beds can be covered with newspaper.

1.8 Handling of spinning larvae and harvesting of cocoons

Once the larvae exhibit spinning behaviour, feeding should be reduced and then stopped. Spinning larvae have to be separated from the bed and moved to the mountages. During the spinning stage, a temperature of 24–25°C and a 60–70% relative humidity should be maintained in the mounting hall. This helps the quality of the cocoons. Adequate aeration should be provided during the spinning period and the spinning larvae should not be exposed to direct heat or sunlight. Cocoons can be harvested after 5–7 days. Dead, flimsy, irregular and stained cocoons should be separated and good cocoons transported to the cocoon market in cotton bags.

1.8.1 Bamboo mountages

Bamboo mountages are the traditional type use in Karnataka. Bamboo strips are fixed in a spiral/circular fashion on the bamboo mats. The ideal size of the mountage is 1.8 × 1.2 m and the gap between the spirals should be 5–6 cm. Some 40–60 larvae per sq ft can be mounted in this manner. During the spinning stage, two mountages placed one behind the other at an angle of 45° reduce the number of bad or rejected cocoons.

1.8.2 Plastic corrugated mountages

Plastic corrugated nets are designed for mounting spinning larvae. The height of the corrugation should be 6 cm and each mountage should have 11 corrugations. The ideal size of the mountage is 60 × 90 cm, and this can be easily placed in a wooden rearing tray; 300–400 nets can be mounted on these mountages. Before mounting the larvae, newspaper should be spread below the mountage which will help to absorb the urination and reduce humidity build up. These types of mountage require less space and can be easily disinfected. Cocoons can be easily harvested. Adequate aeration should be provided in the mounting hall or rack, particularly for CB larvae which display a higher rate of urination.

1.8.3 Rotary mountages

This is a Japanese style of mountage made with thick paper boards. Each mounting frame has 13 rows and 12 columns containing 156 mounting slots. Each slot is 4.5 × 3 × 3 cm in size and the dimension is 55 cm in length, 40 cm in width and 3 cm in depth. Ten such mountages are arranged in a metal frame at a distance of 8 cm from one another. The rotary mounting frames are hung from the ceiling during the spinning stage and can be packed and stored after cocoon harvesting.

Before mounting the larvae, the rotary frames are placed on newspaper. About 1200 larvae are measured and distributed between the rows. Once all the larvae climb onto the mountage, the frames are lifted and hung from the ceiling. Since the spinning larvae exhibit negative-geotrophic behaviour they climb.This weight displacement causes the mountage to rotate, helping the larvae to find an empty slot and settle for spinning. This type of mountage is therefore called a rotary mountage.

1.9 Diseases of silkworms

There are four major diseases of silkworms:

1. pebrine,
2. flacherie,
3. muscardine and
4. grasserie.

1.9.1 Pebrine

Pebrine is one of the most destructive diseases of silkworms and was a major cause of the decline of the silk industry in France in the latter half of the nineteenth century. The name pebrine is so given because of the appearance of the pepper-like spots on the body of silkworms affected by this disease. This is caused by a protozoan parasite called *Nosema bombycis Nageli.* The main spore is oval in shape and is 3–4 × 1.5–2 microns in size. The disease can affect eggs, worms, pupa and even moths. Pebrinized eggs lose sticking power and become easily detached from the egg cards. The eggs lack fertilizing power and die, and the hatchings of even the other eggs are irregular. An important symptom of this disease is the appearance of dark-brown or black spots on the body of the worms (Fig. 1.13). In advanced stages, the larvae become sluggish and dull and show poor appetite resulting in irregular moulting and growth. The worms lack lustre and after the fourth stage the colour becomes rusty brown and the skin becomes wrinkled.

1.13 Pebrine disease.

The only way of preventing hereditary pebrine disease is to use disease-free seed prepared by the Pasteur method. General preventive measures include surface sterilization of disease free layings by dipping egg cards in 2% formalin solution for 20 min and then washing in running water. Other measures include strict sanitation, hygienic rearing, frequent and careful examination of worms for detection of infection, destruction of diseased material, destruction of affected rearing rooms and appliances, and the use of disinfectants such as formalin.

1.9.2 Flacherie

Flacherie is a syndrome associated with bacterial diseases. Diseased/dead silkworms, faecal matter, contaminated mulberry leaves and rearing appliances act as sources of infection. Wide fluctuation in temperature and humidity with poor-quality mulberry leaves are the major predisposing factors for flacherie. The early symptoms of the disease are lethargic larva which stop feeding. At an advanced stage of the infection the larva exhibits retarded growth, vomits gut juices and excretes semi-solid faeces. The larva becomes soft and translucent. Finally the larvae ferments and the inner content turns into a black-coloured liquid which emits a foul odour (Fig. 1.14).

The remedy is to maintain hygienic conditions in the silkworm rearing room by washing the rearing house and equipment with bleaching powder solution followed by disinfectants. It is important to identify and pick out early stage infected larvae and destroy them. If the previous crop has been infected with flacherie then care must be taken to disinfect rearing-room appliances and surroundings before the next rearing. Proper rearing techniques should be adopted by providing quality leaves according to the stage of the worm. During the monsoon season lime should be dusted on the rearing bed to reduce the bed humidity. Depending on the stage of the

1.14 Flacherie affected silkworms.

worms, optimum temperature and relative humidity must be provided and fluctuations of temperature and relative humidity avoided.

1.9.3 Muscardine

Muscardine is a silkworm disease caused by a fungus called *Beauveria bassiana*. The disease prevails mainly in the winter season and often causes massive losses of silkworm resulting in poor cocoon yield. Infection takes place mainly through the integument (skin). The spores (conidia) are airborne and germinate on contact with silkworm's integument. The hyphae then penetrate into the body cavity, grow by sucking the body fluid, invade the tissues and kill the larvae. Low temperature and high humidity in rearing environment favour infection by this fungus. Symptoms of infected larvae are loss of appetite, and sluggishness. Larvae die within 3–5 days of infection. Dead larvae initially appear flabby but harden later due to complete intrusion of the fungus. Finally, the fungus grows over the body surface to produce infectious conidia which may spread further (Fig. 1.15).

Some of the important preventive measures to be taken are: the rearing shed and appliances should be disinfected thoroughly to destroy fungal spores. Silkworms should be reared with adequate spacing, proper ventilation and hygienic conditions. Infected larvae must be collected and burnt before the appearance of conidia to avoid further spread. Manipulation of temperature and humidity in the rearing shed, and the dusting of dry slaked lime powder in the rearing bed are easy preventative measures.

1.9.4 Grasserie

Grasserie is a viral disease in silkworm caused by nuclear polyhedrosis virus (NPV), cytoplasmic polyhedrosis virus (CPV) and infectious flacherie. It is

1.15 Muscardine affected silkworm.

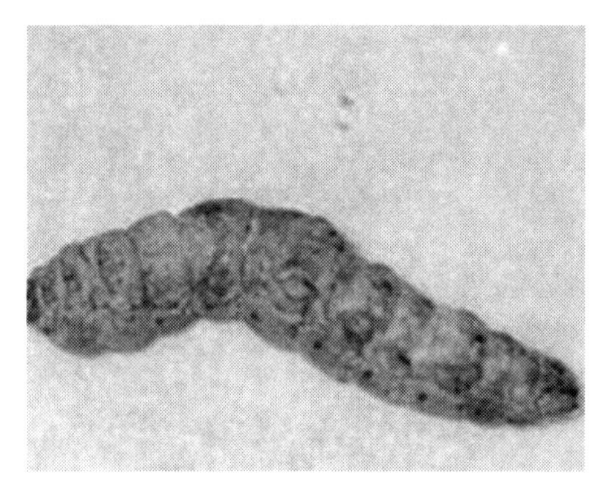

1.16 Grasserie affected silkworm.

caused by the presence of high temperatures, high humidity and the feeding of poor-quality mulberry leaves. It is highly infectious. The symptoms include the presence of polyhedral bodies. As the disease advances, the larvae lose appetite and the skin becomes shiny before moulting. The inter-segment membrane becomes swollen. The haemolymph or body fluid becomes turbid white. Microscopic examination shows the presence of large numbers of polyhedral bodies (Fig. 1.16). The preventive measures for this diseases include, rearing the larvae under clean and hygienic condition, thorough disinfection of rearing room, appliances and surroundings, ensuring proper disinfection of egg surface, incubation of eggs under hygienic conditions, avoiding touching with hands, providing suitable and timely feeding during rearing, maintenance of proper spacing and adequate ventilation. In addition, it is important to pick out and destroy diseased, weak and injured larvae. Application of bed disinfectants as per recommended schedule and quantity may also control this disease.

1.10 Physical characteristics of cocoons

Physical characteristics of cocoons play an important role in deciding their quality and price in the market and during the reeling process. In this section various cocoon characteristics such as colour, shape, size, filament length, shell ratio, reelability, etc, are described in detail.

1.10.1 Colour

Colour is a characteristic particular to the species. Pigments in the sericin layers control the colour. This colour is not permanent and washes away with the sericin during the degumming process. There are diverse hues of colour including but not limited to white, yellow, yellowish green and golden yellow.

1.10.2 Shape

Cocoon shape, as with colour, is peculiar to a given species. Shape can also be affected by the mounting process, especially during the cocoon spinning stage. Generally, the Japanese species is peanut-shaped, the Chinese elliptical, European a longer elliptical and the polyvoltine species spindle-like in appearance. Hybrid cocoons assume a shape midway between the parents, for example a longer ellipsoid or shallowly enclosed peanut form. The shape of cocoons assists in identifying the variety of species and the evaluation of reelability.

1.10.3 Wrinkle

The deflossed cocoon has many wrinkles on its surface. Wrinkles are coarser on the outer layer than within the interior layer. The outline of the wrinkle is not uniform but varies according to species and breeding conditions. Spinning employs high-temperature and low-humidity settings which render fine wrinkles or cotton-like textures of cocoon layers. These provisions discourage the agglutination of the baves resulting from accelerated drying. It is recognized that coarse wrinkled cocoons reel poorly.

1.10.4 Size

Cocoon size or volume is a critical characteristic when evaluating raw materials. The size of the cocoon differs according to silkworm variety, rearing season and harvesting conditions. The number of cocoons per litre, ranging between 60 and 100 in bivoltine species, enables calculation of size. Multivoltine species measure considerably higher.

1.10.5 Cocoon weight

The most significant commercial feature of cocoons is weight. Cocoons are sold in the marketplace based on weight as this index signals the approximate quantity of raw silk that can be reeled. The weight of a single cocoon is influenced by silkworm species, rearing season and harvest conditions. Pure breeds range from 2.2 to 1.5 g, while hybrid breeds weight from 2.5 to 1.8 g. In nature, the weight of a fresh cocoon does not remain constant but instead continuously diminishes until the pupae transforms into a mother and emerges from the cocoon. This weight reduction occurs gradually as moisture evaporates from the body of the pupae and fat is consumed during the metamorphosis process

1.10.6 Thickness/weight of cocoon shell

The thickness of the cocoon shell is not constant and changes according to its three sections. The central constricted part of the cocoon is the thickest segment, while the dimensions of the expanded portions of the head are 80–90% of the central value. The weight of the silk shell is the most consequential factor as this measure forecasts raw silk yield. As with other characteristics introduced in this chapter, shell weight differs according to variety of silkworm. Further, weight is also influenced by the type of technology used for rearing and mounting. In practice, uni and bivoltine species produce heavier shell weights than multivoltine species.

1.10.7 Hardness or compactness

Cocoon hardness correlates to shell texture and is affected by cocoon spinning conditions. For instance, low humidity during the mounting period makes the cocoon layer soft, while high humidity makes it hard. The degree of hardness also influences air and water permeability of the cocoons during boiling. A hard shell typically reduces reelability (during the cocoon reeling process), while a soft shell may multiply raw silk defects. In short, moderate humidity is preferred for good-quality cocoons.

1.10.8 Shell ratio

As the entire cocoon including the pupa is sold as part of the raw material, it is essential to quantify the ratio of the weight of the silk shell vs the weight of the cocoon. This is calculated by the formula:

$$\textbf{Shell ratio}\ \% = \frac{\text{weight of the cocoon shell}}{\text{weight of the whole cocoon}} \times 100$$

This value gives a satisfactory indication of the amount of raw silk that can be reeled from a given quantity of fresh cocoons under transaction. The calculation assists in estimating the raw silk yield of the cocoon and in deriving an appropriate price for the cocoons. The percentage will change based on the breed of the silkworms, rearing and mounting conditions. Percentage rates are altered based on the age of the cocoons as the pupa loses weight as metamorphosis continues. In newly evolved hybrids, recorded percentages are 19–25%, where male cocoons display higher values.

1.10.9 Raw silk percentage

The raw silk percentage is the most important for the value of the cocoon as it has a direct impact on both the market price of the cocoons and the production costs of raw silk. The normal range is 65–84% of the weight of the cocoon shell, and 12–20% of the weight of the whole fresh cocoon.

1.10.10 Filament length

Equally important as the percentage of silk shell is the length of the bave contained within the shell. This factor determines the workload, rate of production, evenness of the silk thread and the dynamometric properties of the output. The length of cocoon filament is a function of silkworm variety. Lengths range between 600 and 1500 m of which 80% is reelable while the remainder is removed as waste.

1.10.11 Reelability

Reelability is defined as the fitness of cocoons for economically feasible reeling. Poor reelability causes a variety of production problems such as halts in production due to filament breakage and high degrees of waste product. Reelability is affected during spinning by cocoon drying, also storage, pre-processing, and reeling machine efficiency and operator skill.

$$\text{Reelability }\% = \frac{\text{number of reeled cocoons}}{\text{number of ends feeding}} \times 100$$

The measured range is from 40% to 80% with deviations depending on the type of cocoon. Note that stained cocoons generally have poor reelability.

1.10.12 Size of cocoon filament

The measure denier expresses the size of the silk thread. A denier is the weight of 450 m length of silk thread divided into 0.05 g units. The diameter of the bave is not constant throughout its length and changes according to its position in the bave shell. At the coarsest section of cocoon filament from 200 to 300 m, the denier increases. These dimensions become finer and finer as the process approaches the inside layer. The average diameter of a cocoon filament is 15–20 microns for the univoltine and bivoltine species.

1.11 Trends in sericulture

New methods of leaf harvesting and advanced rearing technologies have been introduced to obtain quality leaf and compact cocoons with increased silk yield. Some of the latest trends include:

- evolution of appropriate cost-effective technologies through focused research projects addressing constraints and maximizing the production of quality eggs;
- collaboration of sericulture scientists with molecular biologists, bio-engineers, immunologists, textile technologists, clinicians, experts from industry and a host of other stakeholders;
- adoption of region- and season-specific approaches in the development of superior breeds/hybrids and feed package of practices;
- establishment of close linkage between forward and backward sub-systems for greater efficiency and synergy as the sericulture and silk industry is otherwise highly scattered and unorganized;
- identification and promotion of potential clusters for bivoltine and Vanya silk production in potential traditional and non-traditional areas;
- skill upgrading through structured and specially designed training programmes.
- establishment of linkages among the four identified production sub-systems, seed, cocoon, yarn and fabric;
- capacity building for production and supply of adequate quality planting material, silkworm seed, reeling cocoons and silk yarn through promotion of large-scale production units with required techno-financial support;
- development and promotion of participatory extension systems for effective adoption of technologies by similar stakeholders;
- protection to some extent of the Indian silk market from Chinese cheap raw silk and fabrics by implementation of anti-dumping duty;
- effective utilization of by-products for value addition.

1.12 References

Calvert, P. (1985), Biopolymers: the spinning of silk, *Nature*, **315**(2), 17–18.
Calvert, P. (1998), Materials science: silk and sequence, *Nature*, **393**, 309–311.

Chen, Z., Kimura, M., Suzuki, M., Kondo, Y., Hanabusa, K. and Shirai, H. (2003), Synthesis and characterization of new acrylic polymer containing silk protein, *Fiber*, **59**(5), 168–172.

Dhavalikar, R.S. (1962), *Journal of Scientific & Industrial Research*, **21**, 261–263.

Jolly, M.S., Sen, S.K., Sonwalker, T.N. and Prasad, G.K. (1979), Non-mulberry silks. In *Manual on Sericulture*, ed. G. Rangaswami, M.N. Narasimhanna, K. Kashivishwanathan, C.R. Sastri and M.S. Jolly, Food and Agriculture Organization of the United Nations, Rome, 1–178.

Kumar, J.S., Sarkar, A. and Datta, R.K. (2001), A breakthrough in mulberry breeding in sustainable cocoon production. In *Global Silk Scenario – 2001, Proceedings of the International Conference on Sericulture*. Oxford and IBH Publishing Co. Pvt. Ltd., Mysore, India, 242–247.

Manohar Reddy, R. (2009), Innovative and multidirectional applications of natural fibre, silk – a review, *Academic Journal of Entomology*, **2**(2), 71–75.

Matsumoto, A., Kim, H.J., Tsai, I.Y., Wang, X., Cebe, P. and Kaplan, D.L. (2006), *Silk, Hand Book of Fibre Chemistry*, Taylor & Francis Group, LLC., USA.

Mondal, M., Trivedy, K. and Nirmal Kumar, S. (2007), The silk proteins, sericin and fibroin in silkworm, *Bombyx mori* Linn. – a review, *Caspian Journal of Environmental Science*, **5**(2), 63–76.

Rangaswami, G., Narasimhanna, M.N., Kasi Viswanathan, K. and Sastry, C.R. (1976), *Manual on Sericulture Vol. 1*, FAO, Rome.

Robson, R.M. (1998), *Handbook of Fibre Chemistry*, Marcel Dekker, New York, 415.

Sastry, C.R. (1984), Mulberry varieties, exploitation and pathology, *Sericologia*, **24**(3), 333–359.

Sonwalker, T.N. (1969), Investigations on degumming loss and spinning performance of pierced and cut cocoons in silk worm Mori-L, *Indian Journal of Sericulture*, **7**(1), 61.

Spring, C. and Hudson, J. (2002), *Silk in Africa*, University of Washington Press, Seattle.

Ullal, S.R. and Narasimhanna, M.N. (1987), *Handbook of Practical Sericulture*, Central Silk Board, Bangalore, India.

Zarkoob, S., Reneker, D.H., Ertley, D., Eby, R.K. and Hudson, S.D. (2000), U.S. Patent 6,110,590.

Zhou, C.Z., Confalonieri, F., Medina, N., Zivanovic, Y., Esnault, C., Yang, T., Jacquet, M., Janin, J., Duguet, M., Perasso, R. and Li, Z.G. (2000), *Nucleic Acids Research*, **28**, 2413–2419.

2
Silk reeling and silk fabric manufacture

DOI: 10.1533/9781782421580.33

Abstract: This chapter reviews silk reeling and silk fabric manufacture. It discusses types of reeling machine and silk yarns. It also covers weaving technologies and types of woven silk fabric. Finally, the chapter reviews silk spinning.

Key words: silk reeling, silk fabric, silk weaving, spun silk.

2.1 Introduction

Silk is produced by cultured silkworms (Veparia and Kaplan, 2007) and differs widely in composition, structure and properties, depending on the specific source (Robson, 1998; Altman *et al.*, 2003). Post-cocoon technology refers to all activities of silk production after the cocoons have been harvested during the rearing activities. These include:

- cocoon pre-treatment stages (known as silk reeling), such as cocoon sorting, riddling, stifling, cooking, reeling, re-reeling and twisting;
- silk fabric production: fibre preparation, spinning weaving and knitting;
- fabric finishing, including wet processing techniques such as de-gumming, dyeing and printing and
- garment manufacture.

These activities support the ancillary activity of by-product utilization. Silk wastes are used as feed for fish and poultry. Silkworms have recently been used as bio-factories to produce functional protein, which is promoted as a valuable bio-material (Manohar Reddy, 2009).

2.2 Silk reeling

Silk fibres possess outstanding natural properties which rival the most advanced synthetic polymers. However, unlike synthetic polymers, the production of silk does not require harsh processing conditions (Mondal *et al.*, 2007). One of the first operations in silk production is the reeling process

(Sonthisombat and Peter, 2004). Reeling is a vital link in converting the agricultural produce, cocoons, into an industrial output, yarn. Silk reeling consists of a number of activities: cocoon sorting, stifling, cooking, reeling, re-reeling and twisting. Silk reeling basically involves unwinding silk filaments from the cocoons and then reeling the 'baves' (silk fibres complete with their natural gum, sericin), followed by a process in which a number of cocoon baves are reeled together to produce a single thread on a fast moving reel (Das, 1992). Reeling forms a vital link in converting the agricultural produce, i.e. cocoons, into the industrial output of yarn. Raw silk reeling may be classified as:

- direct reeling on a standard sized reel;
- indirect reeling on small reels, and the transfer of reeled silk from small reels onto standard; and
- sized reels on a re-reeling machine.

The last technique is the most commonly used in modern silk reeling processes (Mahadevappa *et al.*, 2001).

2.2.1 Cocoon sorting

This is a process of separating defective cocoons from good cocoons. The process also includes segregating cocoons according to their size. Defective cocoons may be classified as:

- double cocoons;
- pierced cocoons;
- urinated cocoons;
- flimsy cocoons;
- double cocoons;
- pointed or constricted cocoons;
- mould attacked cocoons and
- immature cocoons.

The process of sorting according to size is carried by equipment known as riddling machines. The segregation of uniformly sized cocoons is of great importance as their size influences the cooking and reeling properties.

2.2.2 Cocoon stifling

The main purpose of stifling is to kill the pupa inside the cocoon to avoid its emergence as a moth, thereby preserving the continuity of the filament. This operation also dries the cocoons so they can be stored for a long period. The following methods are generally used in cocoon stifling:

(a) **Sun drying:** The pupa is killed by prolonged exposure (2–3 days depending upon the intensity of sunlight) of freshly harvested cocoons to hot sun. The disadvantage of this process is that continuous exposure to sun hardens the cocoon shell, so affecting the reelability.
(b) **Steam stifling:** The pupa is killed by exposing fresh cocoons to the action of steam for around 25 min. The process can be done by either basket steaming or chamber steaming.
(c) **Hot-air drying:** This is the most effective method and produces good-quality cocoons such as bivoltine varieties. It facilitates the complete drying of cocoons and ease of storage.

Other methods of killing the pupa include the use of infra-red rays, cold air and poisonous gases.

2.2.3 Cocoon cooking

The object of cocoon cooking is to soften the sericin so that the cocoon shell is loosened, enabling the filament to unwind smoothly during reeling. Different methods of cooking may be used, such as open pan cooking, three-pan cooking, pressurized cooking and conveyor cooking.

2.3 Types of silk reeling machines

There are various types of silk reeling machines, including:

- country charka;
- cottage basin;
- multi-end reeling; and
- automatic reeling.

The various types are compared in Table 2.1. The country charka and cottage basin designs are known as sitting type machines.

2.3.1 Country charka

This is a traditional type of reeling machine which provides a simple means of unwinding filaments from the cocoons. Charka units do not require large investment or specific skills and silk produced by these machines is utilized mainly in the hand-loom sector where the cost of raw material has to be

Table 2.1 Comparisons of structural features of various reeling machines

Structural feature	Sitting-type reeling machine	Multi-end reeling machine	Automatic reeling machine
Number of reeling ends per basin	4 (2–8)	20 (10–40)	20
End-groping apparatus	Hand-driven	Semi-automatic	Automatic
End-picking apparatus	Hand-driven	Hand-driven	Automatic
Cocoon supplying apparatus	None	None	Equipped
End-feeding apparatus	None	Equipped (hand feeding)	Equipped (machine feeding)
Stop motion	None	Equipped	Equipped
Traverse guider	Equipped	Equipped	Equipped
Temperature of reeling bath	65–80°C	30–45°C	30–45°C
Reeling velocity (metres per minute)	180–250	50–80	120–200

kept low to make the finished product saleable. Inferior quality multivoltine and defective cocoons can also be reeled more economically on charkas than on cottage basins or multi-end basins (Jolly *et al.*, 1979). Charka silk is generally coarse and displays many defects as no devices such as button/slub catchers and standard croissure systems are used in reeling. Charka silk is not re-reeled. A charka can produce around 1 kg of coarse denier silk per day.

The charka system is an Italian or floating system of reeling. It improves the reelability of inferior and defective cocoons, whatever the quality of silk. It is a simple device, consisting of a large cooking and reeling pan containing boiling water (Fig. 2.1). The cocoons are cooked and the filaments, which are gathered into a bunch after brushing, are passed through a hole in a thread guide device. The thread is then crossed with another thread to form a chambon-type croissure in order to agglutinate the filaments and remove water. It is then passed through a distributor before being wound on to a large wooden reel. Four threads are used in this device. One operator rotates the reel manually and another controls the cocoon cooking and reeling.

2.3.2 Cottage basin

This reeling device (Fig. 2.2) is an improvement on the charka and is designed on the principle of the Japanese multi-end reeling machine. The cocoon cooking is done in a basin of boiling water and the reeling in a hot water basin. Each basin has six ends and each thread is first passed through a button to cleanse it of slubs and waste. It is then passed through

2.1 Country charka.

2.2 Cottage basin reeling machine.

a travellette-type croissure, which is more efficient than that used in the charka. The thread then passes through a traverse guide and finally on to a small reel. As re-reeling is carried out to prepare standard-sized hanks from small reels, the quality of silk produced on this machine is superior to charka silk. A basin can produce around 800 g of silk per day and is capable of reeling superior quality cocoons such as bivoltine.

2.3.3 Multi-end reeling machine

This machine eliminates the disadvantages of the sitting-type reeling machine by increasing the number of reeling thread ends per basin and reducing the reeling speed. The operator stands when running this machine as the number of reeling threads per basin is increased 20-fold. It is therefore also known as a 'standing-type reeling machine'. The reeling efficiency is unchanged and the quality is better due to the reduced speed. The multi-end reeling machine is composed of a driving part, groping ends, picking ends, standby bath, reeling part, jetboute, stop motion, traverse guide, small reels, steam heating pipes and clutches.

The cooked cocoons contained in the tubs are carried into the groping ends portion of the reeling machine. From there, the cocoons are moved into the picking ends apparatus. After processing, the cocoons go to the standby bath where they are picked up by the reeler and fed to the reeling thread. During this step a number of cocoons will be dropped, thus reducing the ratio of reeling cocoons per thread. A skilled reeler will feed around 16 cocoons per minute. The reeling thread passes through a jetboute, silk button, four guiders and a traverse guide, and then is wound onto small reels (Lee, 1999).

Cocoons dropped during the reeling process are gathered and reprocessed, starting from the groping end section. The croissure of the reeling thread is made between the second and third guiders and the thread is given cohesion by rubbing the combined filaments rather than by twisting. A set of multi-end reeling machines typically consists of ten basins, each basin having 20 ends or reels.

The key components are:

- **Basin:** The basin is rectangular with well-rounded corners and edges, 10–12 cm deep and is commonly made of dark-coloured porcelain. The basin is subdivided into sections, each of which is intended for a specific task such as brushing, end gathering of the baves, reserve stocks and waste collection.
- **Reels:** Reels in a multi-end reeling machine have a circumference of 75 cm. The frame is made of light metal or plastic and the reels are fitted into carriers which are driven by a transmission shaft through connecting gears.
- **Traverse guide:** To ensure a long narrow web on the hank of the reel, a cam-type traverse assembly is fixed. This gives the hank a convex surface as it is wound on the reel. The centre part of the hank is therefore higher than the two axes.
- **Thread button:** Porcelain button thread-guiders are used for removing any dirt adhering to the thread which passes through a very small aperture in the button.

The multi-end reeling machine (Fig. 2.3) is a modern reeling device which gives a better performance with superior quality cocoons such as bivoltines.

2.3 Multi-end reeling machine.

2.3.4 Automatic reeling machine

The continuing rise in labour costs has made automation necessary in the production of raw silk. The automatic reeling machine (Fig. 2.4), which controls the number of reeling cocoons per thread, was invented around 1950. This was soon replaced by a machine which could automatically control the size of the reeling thread. An automatic reeling machine usually has 400 ends, while one basin has 20 ends. The operating efficiency of automatic reeling machines may be affected by cocoon quality, drying and cooking machinery and the quality of the reeling water.

The automatic reeling machine mechanizes the processes of groping ends, picking ends, cocoon feeding to reeling thread and the separation of dropped cocoons during the reeling process (Sonwalker and Krishnaswamy, 1980). Its efficiency compares favourably with that of the manual multi-end reeling machine. Although designed to replace manual reeling, the automatic reeling machine still requires the manual correction of problems which may occur with the reeling thread.

A moderate number of cooked cocoons are carried to the newly cooked cocoon feeder and are then moved into the groping end section. They are then moved to the picking end section and dispensed to the cocoon supplying basket which rotates continuously around the reeling basin on a chain belt. The reeling method is classified into:

- fixed cocoon feeding system and
- moving cocoon-feeding system.

2.4 Automatic reeling machine.

In the case of the fixed cocoon feeding system, the correctly picked end cocoons in the rotating baskets are poured into the arranging basins where the picked end of each cocoon is hung on the end holding reel. When the size detector for the reeling thread so indicates, the waiting cocoons are passed to the reeling thread by a feeding spoon. The reeling thread then passes through a jetboute, silk button, four guiders, a denier indicator, a fifth guider and a traverser, and is finally wound onto small reels. The end dropped cocoons are placed into the cocoon transporting channel by a remover plate and are carried into the pupa separating drum. Further reelable cocoons are delivered into the end-groping section by the conveyor belt and the reel-finished cocoon placed in the dropped-pupa case for parchment layer cocoons.

In moving cocoon feeding systems, the correctly picked end cocoons are placed in a moving basket equipped with a feeding apparatus which moves around the reeling basin. The denier indicator for the reeling thread determines the feeding motion of the cocoon. After cocoon feeding, the reeling path of the moving cocoon feeding system follows that of the fixed feeding system.

2.4 Types of silk yarn

The range of yarn types produced by reeling is as follows:

- **Poil:** A silk yarn formed by twisting raw silk. The twist may be very slight or may exceed 3000/m.
- **Tram:** A silk yarn formed by doubling two or more silk threads and then twisting them slightly, usually between 80 and 150 turns per meter

(TPM). The resultant yarn helps to give body to a fabric while displaying the characteristics of silk such as sheen and smoothness, although it lacks the strength of organzine. These threads usually have 3–6 turns per inch, depending upon the kind of fabric for which they are intended.

- **Crepe:** A silk yarn made by doubling several raw silk threads and twisting them in the range of 2000–4000 TPM. The threads, especially the tram, need special treatment to produce a crepe. The single threads are put together as in a regular tram, but are twisted very tightly. Instead of two or three turns an inch as in tram, the yarn is given between 40 and 80 turns per inch. When woven into fabric, the elasticity of these threads causes them to 'kick up' or crinkle, so creating the crepe effect. Crepe yarn is used in making crepe de chine, crepe charmeuse, crepe meteor, crepe faille, crepe organzine in satin charmeuse and chiffons, as well as other fabrics. Crepe yarn may be variously combined with regular yarns and different weaves, each variation producing a different effect in the appearance and texture.
- **Organzine:** Organzine is prepared by twisting a single thread and then combining it with other twisted singles, the number depending upon the size of thread required. The threads are then combined by twisting in the direction opposing that given to the single threads For example, if the singles are twisted by a turn to the right, the combined singles are given a left-hand twist. The result is a hard-finished, smooth, strong thread, comparatively small in diameter. Completed thread which is ready for the loom usually has between 10 and 14 turns to the inch.
- **Grenadine:** A silk thread formed by doubling two or more ends of poil and twisting them in the opposite direction to that of the individual poil ends. Grenadine is three to four times more tightly twisted.
- **Cordonnet:** A thick silk thread obtained by doubling and a process known as throwing, which winds several tram ends in the opposing direction to the twist of the individual tram ends.

2.5 Key steps in silk fabric manufacture

There are various stages in silk manufacture which include:

- soaking;
- winding;
- doubling;
- twisting;
- warping and
- pirn winding.

A flow chart representing the stages in the manufacture of silk fabric is shown in Fig. 2.5.

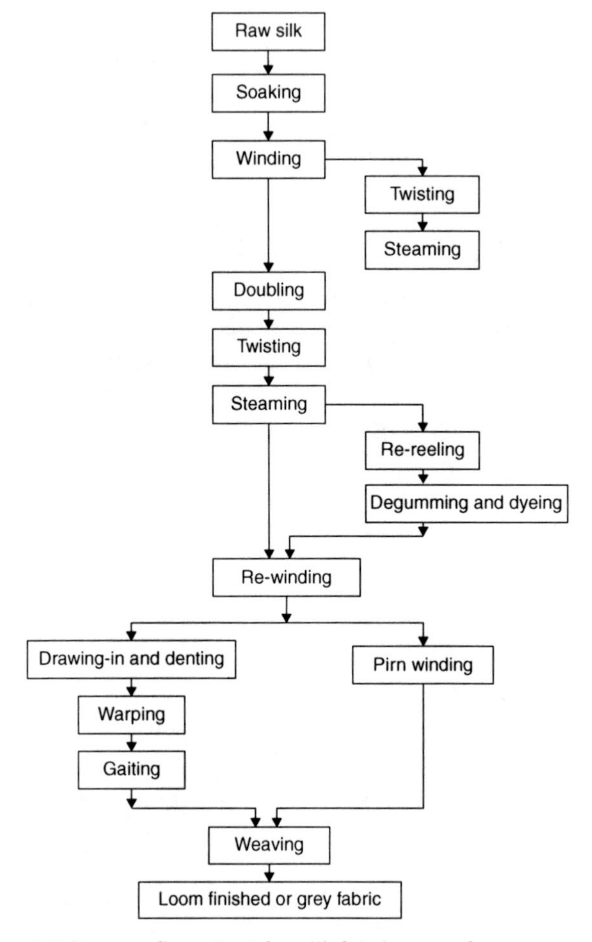

2.5 Process flow chart for silk fabric manufacture.

2.5.1 Soaking of silk

The raw silk hanks intended for weaving are first subjected to a soaking process. The main reasons for soaking are:

- to soften and slacken the gum and
- to lubricate the thread and make it more pliable.

The materials used for soaking are soap, oil, water and occasionally salt. Soap is used to soften the thread and oil to lubricate it. The water used for soaking should be neutral and soft. Borax is a popular solvent used in the soaking process which typically lasts 12–14 h at a temperature around 40°C. The following recipe is recommended for soaking:

- soap: 3%;
- oil: 4%;
- water: 1:9 and
- borax: in very small quantities.

2.5.2 Winding

Winding is the process of putting the raw silk onto bobbins in the form of hanks. Hard gum spots, loose ends and knots are removed in the winding operation. Double-flanged bobbins are used in winding machines which consist of 50–100 bobbin units. Winding should be carried out at 65% ± 2% RH. The daily production per drum is around 100–150 g.

2.5.3 Doubling

In this process, two or more threads from the winding bobbins are placed parallel and wound on to bobbins to facilitate twisting them together. Doubling reduces irregularities in the yarn and improves the strength. A doubling machine is similar to a winding machine. Winding bobbins are placed on the creel and the yarn is passed through an individual thread stop motion device, tensioner and thread guide and on to a double flanged bobbin which is mounted horizontally. The doubling tension should be in the range of 0.2–0.3 g/denier. The cover factor of fabrics is improved where doubled yarns are used. Doubling takes place in 2, 3, 4 ply or greater. The machine capacity is 50–100 double-sided bobbins and the production capacity is two to four times that of the winding machine.

2.5.4 Twisting

Silk is rarely used directly for weaving. It first undergoes a process of twisting which depends upon the type of fabric to be produced. The strands are wound together to create a silk yarn which varies in weight and texture depending on the number of strands and the twist (Spring, 2002). The twisting machines used for silk are of a type known as uptwisters. The doubling bobbins are placed on vertical spindles which rotate at speeds of about 8000–10 000 rpm. The twist applied to the thread depends upon the relation between the speed of rotation and the winding speed of the take up roller:

$$\text{TPM} = \frac{\text{Spindle speed (rpm)}}{\text{Front roller speed (mpm)}}$$

where rpm is revolutions per minute and mpm is metres per minute.

Twisting affects the brilliancy of a thread. As the twist increases, the lustre of the thread decreases and it becomes opaque.

Twisting machines normally have two or three rows of spindles in order to save space and are usually double sided. The degree of twist in the silk yarn depends upon the end use of the fabric: yarns such as crepes and georgettes require a high rate of twisting. The daily production of a twisting machine/spindle is about 25–30 g for two-ply (20/22 d) yarns. The twisted yarn is subjected to steaming in an autoclave which fixes the twist by making the thread more ductile. The duration of steaming depends on the depth of the layers of thread on the bobbins and their level of twisting. Ten minutes is usually sufficient for organzines, while more highly twisted crepes require around 2 h.

2.5.5 Warping

Warping is the process of collecting individual ends from a creel and transferring them to a beam. Depending on the density of warp ends and the width of the fabric, the number of ends required for the fabric is divided and the number of sections decided upon. Three different methods of warping are used, sectional, beam and ball warping. Sectional warping is preferred for silk warp because of the fine denier and higher number of ends required. Ball warping is widely used in the hand-loom weaving sector.

2.5.6 Pirn winding

Pirn winding is necessary in the preparation of weft yarn and pirn winding machines may be automatic or hand operated. The pirns used in power looms are usually larger and the yarn content greater than in handlooms where pirns are smaller and a hand-operated charka is used for their preparation.

2.6 Weaving of silk fabrics

Silk is a delicate fabric which is valued in fashion and commands a high price. Its production is a long and laborious process requiring great attention to detail and was kept secret in China until about AD 300 when the process was leaked to India. Although the process of weaving silk has undergone many changes over the centuries, the core method has remained the same. Silk seems to have played an important role in the development of loom and weaving technology. Traces of primitive looms and woven fabrics have been found in excavations in Egypt, China, India and Peru. These two-bar bamboo devices were later improved to include horizontal and vertical shaft looms but were suitable only for plain or simple patterned coarse weaving, carpets, tapestry or floor coverings. The silk weavers of China were the first to use the heddle and draw loom, a revolutionary development over the traditional primitive loom. Another technical innovation came from India with the invention of a foot treadle for silk weaving (Datta and Nanavaty, 2005).

Silk was originally woven by hand and that method continues to the present day. However, with the introduction of a variety of looms and tools, the process has become easier and faster. Modern handlooms still use the same process of laying out the warp threads onto horizontal beams and weaving with the use of a heddle and foot pedal. The Jacquard loom was the first automated loom to lace and feed weave designs into the loom. Today, the plain loom is the most commonly machine for one-colour fabrics. It is power-operated and uses only one shuttle. The most advanced apparatus is the shuttle-less loom in which threads are inserted into the loom from revolving cones and then drawn across by a long needle or by powerful, narrow jets of water.

Specializations are necessary in the weaving of silk goods to achieve the required fabric standards. Because of the high denier variations in the yarn, the predominance of weft bars may be reduced by using a box mechanism to mix the weft yarn. The improvement in fabric quality compensates for the reduced productivity of box looms. The use of a dobby in silk weaving is a minimum requirement because the fineness of the silk ends means that a high number are required.

Fitting a large number of ends within heald shafts creates problems. However, the use of underpick looms provides optimum weaving conditions and permits the weft to be mixed by the use of box mechanisms. As silk filaments are susceptible to variations in tension, long distances between the reed and the back rest are necessary, as are abrasion-free surfaces on the race board and smooth shuttle surfaces. The use of a rotary back rest will reduce warp breakages and rubber covered take up rollers and porcelain eyes set in heald wires should be used.

In order to manufacture a quality fabric in widths acceptable to international markets, it is desirable for silk-weaving looms to be standardized. A standard silk loom should be of the underpick type, having a box mechanism with metallic reed, dobby or jacquard and a highly polished (chromium-plated) rotating back rest with warp and weft stop motions. Superior quality yarns should be used as the warp, while a slightly inferior grade may be used for the weft. It is not advisable to reduce the fabric weight by reducing the number of ends at the cost of density of the weave. It is always better to use finer deniers to weave fabrics of lower weight, so ensuring a firm weave and stable cloth construction.

Silk weaving is traditional in most countries. It is largely carried out on handlooms, power looms providing only a small portion of the total production. Most handlooms are of the fly-shuttle type and each locality has its own tradition. The looms used for specialized weaving, such as tie and dye and balucher, operate on throw shuttles. The different types of handlooms include the pit loom (Fig. 2.6) and the frame loom.

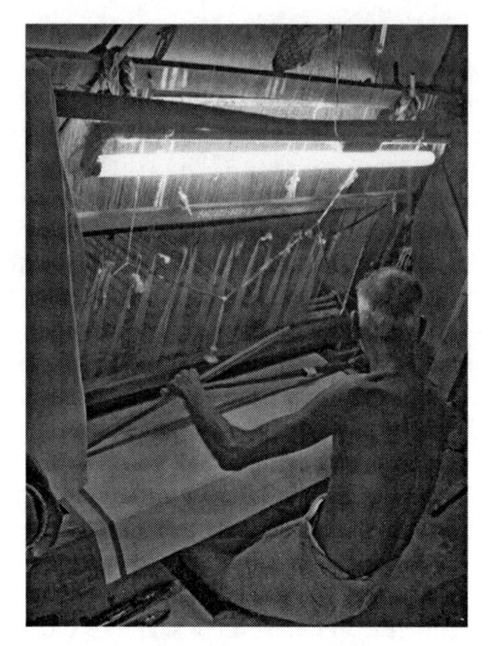

2.6 Pit loom for weaving silk fabrics.

2.7 High-speed rapier loom for weaving silk fabrics.

Weaving with modern high-speed machines is efficient, provided there are minimal stoppages. Due to the relatively high investment costs, every stoppage represents a loss of productivity and profit. It is therefore essential to use a quality of yarn for warp and filling which will meet the requirements of high-speed machines. Warp yarn is exposed to much higher stresses than the weft yarn and there are a variety of factors which contribute to warp breaks. The strength of the yarn is important as are the number of knots, the cleanliness and cohesion. Due to the high speed at which filling yarn is inserted, the quality of the cone is also of great importance. Any disturbance or inertia in unwinding may stop the loom. Uneven thickness in the case of spun silk is also critical for weaving on these looms (Fig. 2.7).

2.7 Types of silk fabric

There are various types of silk fabric, including:

- charmeuse;
- crepe de Chine (CDC);
- georgette and
- habutai.

2.7.1 Charmeuse

Charmeuse is the most common type of silk fabric. It is well known in clothing markets and is usually characterized 'by its lustrous shine and sumptuous feel and is primarily used for the manufacture of skirts, dresses, evening wear, nightgowns, lingerie, and gently shaped tops', according to Silktrading.com. Charmeuse does not unravel easily and is fairly difficult to sew.

2.7.2 Crepe de Chine (CDC)

Crepe de Chine (also known as CDC) is a very durable, lightweight and wrinkle-resistant silk fabric with a dull surface and a 'pebbled texture', according to Silktrading.com, who also describe it as 'a hot favourite among the designers and is primarily used for fashionable and sophisticated skirts, dresses, suits and evening wear', CDC is categorized as a mulberry silk fibre.

2.7.3 Georgette

Georgette silk fabric is similar to crepe de Chine. 'It is soft and lustrous, drapes very easily and falls into soft ripples. It is characterized by a "grainy" texture and is used for the manufacture of dresses, skirts, blouses, tops and evening wear', according to Silktrading.com. Georgette is strong but can snag easily. It is more difficult to sew and repair and may require a sewing machine with a special foot. It does not show pin marks and is generally wrinkle-free.

2.7.4 Habutai

Habutai means 'soft and downy' in Japanese, according to Silktrading.com. This type of silk cloth was originally used to make kimonos. It is a lightweight, shimmering material with a soft feel and is commonly used for making garments such as skirts, suits, lingerie, dresses and evening wear. Habutai is typically sheer and has a natural ivory colour.

2.8 Spun silk production

A certain amount of waste is generated in all sectors of the silk industry. The waste is of different types and although a large amount of cocoon fibre is unfit for reeling or working up into the finest grade of silk fabrics, it can be spun and used in the manufacture of goods which differ in grade from the highest priced fabrics. An average of 35% by weight of silk waste from raw silk reeling is being produced in the reeling industry both from the mulberry and non-mulberry sectors (Sonwalker, 1993).

During the process of mulberry silk reeling the different qualities of silk obtained as by-products are (Sonwalker, 1993):

- **waste from cocoons:** floss, spelaia (Italian), discarded cocoons, pierced cocoons, double, stained cocoons, etc.;
- **reeling waste**: cooker's waste, reelers' waste, basin re-use or boiling off waste and
- **thread waste:** re-reeling, winding or throwers' waste, weaving waste (hard waste).

Methods of utilizing silk wastes vary considerably, according to the character of the waste. Raw waste from cocoons needs a different treatment from that given to silk shoddy. Short-fibred waste is treated differently from long-fibred waste, while-coloured waste goes through processes which are not used for white or uncoloured wastes. The first step in manufacture using silk wastes is the classification of the material into as many varieties as practicable. Some sorting and classifying is done by the producers of the raw silk and some by silk rag dealers who collect the waste, but most of the sorting is done within silk-manufacturing units.

Silk waste is normally collected by trading agencies from silk filatures and supplied to spun-silk-manufacturing units. A spun-silk mill may be broadly divided into five main divisions:

1. de-gumming;
2. dressing or combing;

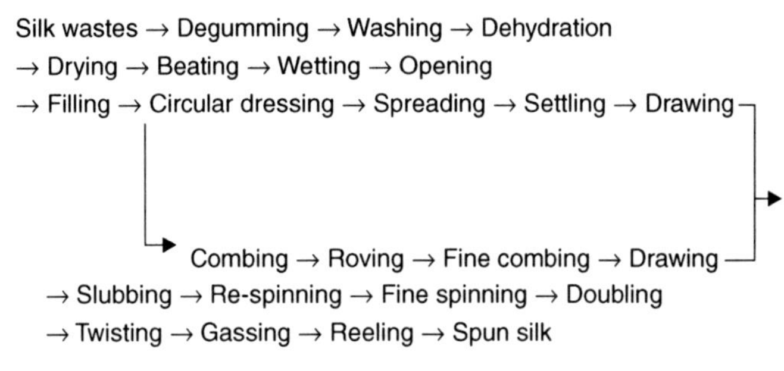

2.8 Spun silk manufacturing process

3. preparatory operations;
4. spinning and
5. finishing.

A schematic diagram showing spun-silk manufacture is presented in Fig. 2.8.

2.8.1 De-gumming of waste silk

The first step in treating raw silk waste is to prepare the fibres for subsequent mechanical processing by removing the gum or sericin. There are two methods:

1. soap and soda boil method and
2. enzymatic de-gumming.

The soap and soda boil method involves treating different qualities of waste silk with slightly alkaline soap solution for around 1 h. The liquor is heated from the bottom by steam coils. The material is worked either manually or mechanically in the vat where the liquor ratio is about 1:30, 25% soda is prepared and the material is processed for 45 min, after which the liquor is discharged. A second boil is carried out with a 25% soap solution and washed twice in the same vat. The de-gummed material is then hydro-extracted to remove water and is then dried, either in the open air or in a heated closed chamber according to the season.

2.8.2 Dressing or combing

De-gummed waste is processed in a machine known as an opener which transforms the material from a tangled mass of fibre into a ribbon in which the filaments will be more or less parallel. The opener has two small diameter feeding rollers with small metallic teeth and another drum of larger diameter. The material is fed to the roller on a belt and passed to the teeth of the rotating drum. The fibres deposited on the drum lie largely parallel and are removed when they reach a certain weight.

The drum of the combing or circular dressing frame has a diameter of 1.7 m and turns slowly. The combing mechanism consists of two rollers studded with metallic points as wide as the drum, but of much smaller diameter, one at each side of the lower part of the drum and turning in opposite directions so that the material is combed first in one direction and then in the other.

2.8.3 Preparatory operations

After the dressing process, the first preparatory stage is cleaning the slivers. These are inspected on a glass panel, defects being corrected and any remaining foreign matter removed by hand. The key steps are:

- **Spreading:** The spreader first combines the slivers and then gradually draws out a ribbon until it is of an even size. The degree of stretching by the spreader is from 1:6 to 1:12 (draft) according to the quality of the material. The ribbon is gathered on the surface of a large drum and then proceeds to a series of drawing operations.
- **Drafting:** The material is fed into the rear set of rollers then delivered to the front set. The rate of delivery always is faster than the rate of feed. Supposing the front roller delivery (surface speed) to be ten times the surface speed of back rollers, then the draft would be ten. If the fibres are fine and long, the drafting is easier and thus higher drafts can be employed than when processing short and coarse fibres.
- **Drawing:** This series of preparatory operations takes the ribbon or slivers through six different drawing machines, in each of which it is submitted to extension varying from 1:6 to 1:10.
- **Roving:** The preparatory stages are concluded with a final extension of about 1:8. The completed yarn or roving, in which the final operation imparts a slight twist, is now ready to be spun on a ring spindle machine.
- **Twisting:** As drafted material from the roving stage becomes finer, there will be a smaller number of fibres in the cross-section and the fibres will be laid parallel. In order to bind the fibres in a manner which will give sufficient strength to the material, twisting is necessary and to select the appropriate amount of twist for a particular material, the following constants are used:

  – TPI (twist per inch) = twist factor × $\sqrt{\text{count}}$

– also, $\text{TPI} = \dfrac{\text{Spindle speed}\,(\mathbf{rpm})}{\text{Front roller delivery in inches}}$

where rpm = revolutions per minute

– Production/spindle for 8 h (g): $= \dfrac{\text{FRD in inches} \times 60 \times 8 \times 453.6}{36 \times 840 \times \text{Cotton English count}}$

where FRD = front roller diameter.

2.8.4 Spinning

In ring spinning, a spindle surrounded by a steel wheel turns a small metal ring, known as a traveller, which is attached to it. The spindle rotates at a speed of about 6000–7000 rpm. The yarn is drafted to an extent of 1:15 or 1:30 as it passes through the traveller, which is also rotating rapidly. There are three operations in ring spinning, which take place simultaneously:

1. **drafting** – carried out on a set of rollers known as drafting rollers;
2. **twisting** – caused by the difference in speed between the spindle or traveller and the front delivery roller and
3. **winding** – caused by the speed of the traveller lagging slightly behind that of the spindle due to yarn drag.

2.8.5 Finishing

There are a number of processes involved in finishing, including:

- **Doubling:** Two or more threads are combined to make weaving or sewing threads respectively. After doubling, a second twist is imparted by the throwing machine, the number of turns varying with the property required of the thread.
- **Gassing:** The twisted yarn is then moved to the cleaning and gassing frame where it passes through a gas flame under friction obtained by winding the yarn round a number of runners. It is passed repeatedly through the flame at a speed of 500–600 mpm. The friction imparted by the runners helps to eliminate neps, other impurities and weak areas in the thread, while the flame burns away protruding fibres and imparts a gloss to the silk. Any adhering burnt particles are removed from the thread by passing it between rotating steel rollers in a process termed as cleaning. These gassing and cleaning operations are performed by a single machine.

- **Reeling:** The gassed yarn is then reeled onto frames (straight or cross reel) to form a hank. The hanks are dressed and folded in bundles which are press packed into bales ready for the market. The usual counts (metric) of yarn spun in Indian silk mills are 60, 140 and 210 s (two-fold).

2.8.6 Uses of spun silk yarns

Silk waste or floss is used in making spun-silk yarns for the manufacture of lining silk, knit goods, hosiery, mufflers, cheap silk neckties, coarser qualities of sewing thread, pile fabrics, elastic webbing and certain kinds of dress goods, and are mixed with wool for particular effects. Spun silk is also used in the manufacture of lace and embroidery but is less fine and strong than reeled silk (Gangopadhyay, 2008). The best grades of spun silk yarn are used as filling or weft in several varieties of silk fabrics, both plain and twill, and in pile goods such as velvets. High-grade spun-silk yarn is also used as warp in goods that have a cotton or wool filling. A considerable amount is also used in the production of embroidery and knitting silks. Lower grades of spun-silk yarn are used in making ribbons and silk cords, while the cheapest grades are used to make knitted goods and coarser silk or silk-mixed fabrics. The poorest grades of spun silk, which are carded rather than combed, are used as filling in cheaper grades of silk dress goods, silk upholstery fabrics, polishing cloths and coarse grades of knitted goods.

2.9 Future trends in silk production

Silk is considered to be one of the finest natural fibres in the world. It exhibits properties such as great tensile strength and extensibility combined with very good comfort and dyeing properties. The fibre is important in a global context as it is the only natural fibre available in continuous filament form. It is synonymous with luxury, beauty, sensuality and elegance, and possesses qualities unrivalled by any other fibre. It is light, strong, smooth, soft and versatile, and being a natural fibre it is more comfortable to the skin. Silk used to be perceived as a luxury fibre, but in the early 1990s, it became more affordable due to the rising popularity of machine washable sand-washed silks.

Silk accounts for less than 1% of the world fibre market by volume but its importance in terms of value is far greater. Exports of silk goods from major silk-producing countries such as China have exceeded more than US$ 3 billion in recent years. The USA has been a major importer of silk from India and China. India is the second largest producer of silk and has considerable potential for improving its production capacity. Product development and diversification will be of great importance if India is to harness its inherent strengths towards meeting market requirements. The export market base will have to be widened and the range of products broadly based to meet customer expectations.

The domestic market for silk is substantial. Product diversification into casual wear and alternative end uses, with a focus on the growing upper middle-income group, will add value to domestic silk production in India. In recent years, increased awareness of the non-mulberry, or so-called 'vanya', silk products in the domestic markets of China and India has been evident, especially in regard to tasar and eri silk. Product development and diversification in tasar, eri and muga has been largely an unexplored area in India, although the importance of new designs, motifs and products has now been recognized in the vanya sector. This has given impetus to product development which will contribute to diversification and enlargement of the market share. However, mulberry silk still constitutes the bulk of production in India and China and is being developed with a growing clientèle and enlarged product range in mind.

New trends in the production and applications of silk are being deployed to maintain a sustainable market. The world's first silks exhibiting fluorescence and other novel properties have been successfully developed as a result of transgenic silkworm research conducted by Japanese researchers who have now developed three lines of transgenic silkworms. The first line produces silk threads emitting green, re or orange fluorescent light which are created by introducing genes which promote the generation of fluorescent proteins into the silkworm eggs. The second produces green fluorescence by the introduction of genes extracted from jellyfish, a technique developed by Nobel Prize-winning chemist Osamu Shimomura. The third line uses genes extracted from coral, a technique that has already been used in commercial applications to produce red and orange fluorescence. Fluorescent silk threads have great potential for use in the fashion industry and it is expected that there will be considerable demand for them from producers of high-end apparel.

Though silk is produced in more than 20 countries, the major producers are in Asia. Sericulture industries have also been established in Brazil, Bulgaria, Egypt and Madagascar where their labour-intensive advantage makes them attractive. Silk is in demand for its compatibility, eco-friendly and value-added potential, for its nutritive value in the diet of cardiac and diabetic patients, and as a component for cosmetic preparations. Bio-compatibility has made silk a base material for the reconstruction of tissue walls, membranes, muscle ligaments, blood vessels, nerves, cartilage and bone.

The future of silk in the twenty-first century is bright. Though consumption is likely to rise despite its high price, it remains to be seen whether silk production will keep pace with rising demand. Japan is the highest consumer of silk and the growth in Chinese standards of living is boosting the demand for silk within the country. Almost 85% of the increase in Indian silk production is consumed within the country which currently imports more than 5000 MT every year to meet the shortfall. Ecological factors are also increasing the demand for silk in Europe and the USA.

Traditionally, silk has been produced solely for textiles. However, new approaches have extended its application into the areas of nutrition, cosmetics, pharmaceuticals, biomaterials, biomedical and bioengineering, automobile manufacture, house building, art and crafts. 'Smart' clothing and innovatory performance garments offer further opportunities in silk technology and design. Innovative textiles enriched with sense experiences other than the visual are in the offing. The silk world may be revolutionized in the coming decades by research and development producing sense-enriched, fragrant, cooling and mosquito-repellent silks

The increasing global demand for variable eco-friendly silk composites and the consequent impact on value, employment and environmental safety will require greater awareness among stakeholders and trainers, growth in the exchange of ideas between entrepreneurs and accessibility for consumers. The International Year of Natural Fibres in 2009 created a greater awareness of innovative research into silk and its future uses.

2.10 References and further reading

Altman, G.H., Diaz, F., Jakuba, C., Calabro, T., Horan, R.L., Chen, J., Lu, H., Richmond, J. and Kaplan, D.L. (2003), Silk-based biomaterials, *Biomaterials*, **24**, 401–416.

Das, S. (1992), *Indian Textile Journal*, **102**(12), 42–46.

Datta, R.K. and Nanavaty, M. (2005), *Global Silk Industry: a Complete Source Book*, Universal Publishers, Boca Raton, FL, USA, 171.

Gangopadhyay, D. (2008), Sericulture industry in India – a review, *India, Science and Technology*, 13–18.

Jolly, M.S., Sen, S.K., Sonwalker, T.N. and Prasad, G.K. (1979), Non-mulberry silks. In *Manual on Sericulture*, ed. G. Rangaswami, M.N. Narasimhanna, K. Kashivishwanathan, C.R. Sastri and M.S. Jolly, Food and Agriculture Organization of the United Nations, Viale delle Terme di Caracalla, Rome, 1–178.

Lee, Y.-W. (1999), *Silk Reeling and Testing Manual*, FAO Agricutural Services Bulletin No. 136, Food and Agriculture Organization of the United Nations, Rome.

Mahadevappa, D., Halliyal, V.G., Shankar, D.G. and Bhandiwad, R. (2001), *Mulberry Silk Reeling Technology*, Oxford and IBH publication Co., New Delhi and Kolkatta, India.

Manohar Reddy, R. (2009), Innovative and multidirectional applications of natural fibre, silk – a review, *Academic Journal of Entomology*, **2**(2), 71–75.

Matsumoto, A., Kim, H.J., Tsai, I.Y., Wang, X., Cebe, P. and Kaplan, D.L. (2006), Silk. In *Hand Book of Fibre Chemistry*, ed. M. Lewin, Taylor & Francis, New York.

Mondal, M., Trivedy, K. and Nirmal Kumar, S. (2007), *Caspian Journal of Environmental Science*, **5**(2), 63–76.

Robson, R.M. (1998), *Handbook of Fibre Chemistry*, Marcel Dekker, New York, 415.

Sonthisombat, A. and Speakman, P.T. (2004), *Silk: Queen of Fibres – The Concise Story*., Department of Textile Engineering, Faculty of Engineering, Rajamangala University of Technology Thanyaburi (RMUTT).

Sonwalker, T.N. (1993), *Hand Book of Silk Technology*, Wiley Eastern Limited, New Delhi, 181.

Sonwalker, T.N. and Krishnaswamy, S. (1980), Working of an automatic reeling machine in CSR&TI, Mysore, *Indian Silk*, **November**, 18–20.

Veparia, C. and Kaplan, D.L. (2007), Silk as a biomaterial, *Progress in Polymer Science*, **32**, 991–1007.

3
Structural aspects of silk

DOI: 10.1533/9781782421580.56

Abstract: This chapter reviews the structure of silk. Topics include chemical structure, amino acid composition, moisture regain properties, microstructure and crystalline structure, as well as optical properties.

Key words: amino acid composition of silk, moisture regain, microstructure of silk, optical properties of silk.

3.1 Introduction

Silks belong to a group of high molecular weight organic polymers which are characterized by repetitive hydrophobic and hydrophilic peptide sequences (Altman *et al.,* 2003). They are protein polymers which are spun into fibres by some arthropods such as silkworms, spiders, scorpions, mites and fleas (Craig, 1997; Altman *et al.,* 2003). There are thousands of silk-spinning insects and spiders, such as *Nephila clavipes* and *Bombyx mori* (Bell *et al.*, 2002; Becker *et al.*, 2003), although only a few have been investigated in detail. Silks differ in composition, structure and properties depending on their specific source and function (Craig *et al.*, 1999; Altman *et al.,* 2003; Sheu *et al.*, 2004). Silk produced by spiders is mechanically superior to any insect silk. However, the silk produced by mulberry silk-worms is superior in other respects and is the main source of manufactured silk.

Silkworm fibres are classified as domestic silks and wild silks. Wild silks are produced by caterpillars other than the mulberry silkworm and they differ from the domesticated varieties in colour, size and texture. Cocoons gathered in the wild usually have been damaged by the emerging moth, so the silk thread that makes up the cocoon has been torn into shorter lengths. Domestic silkworms like *B. mori* are commercially reared and the pupae are killed by dipping them in boiling water before the adult moths emerge, so allowing the whole cocoon to be unravelled as one continuous thread. There are other commercially exploited silkworms which are known as non-mulberry silkworms, such as *Antheraea mylitta* (tasar), *Phylisamia ricini* (eri) and *Antheraea assama* (muga).

3.2 Composition of silk

In contrast to all other natural fibres, silk does not have a cellular structure. In this respect, and in the way it is formed, it closely resembles a man-made fibre. Morphologically, silk is very simple. It consists of two single compact, continuous threads which are extruded by the silkworm as it spins its cocoon. These are surrounded and covered by silk gum or sericin. The denier of the filament varies within the cocoon.

A silkworm extrudes liquid fibre from the two excretory canals of sericteries which unite in the spinneret in its head. Each of these two threads is known as a brin. The two brins are cemented together in the spinneret by sericin to become a single continuous fibre called the bave or filament (Carboni, 1952). Sericin acts as a glue and fixes the fibroin fibres together in a cocoon. Sericin and fibroin protein has useful properties and has been found to possess various biological functions. Genetically, silks are characterized by a combination of highly repetitive primary sequences which have significance in the secondary structure and provide unique mechanical properties. These properties, combined with the biocompatibility, of silks, have led to their use in controlled release and other types of biomaterial in addition to their uses as textile materials (Yao and Asakura, 2004).

The molecular weight of sericin ranges from 10 to 310 kDa and fibroin ranges from 300 to 450 kDa (Tanaka *et al.*, 1999; Zhou *et al.*, 2000) and the key amino acids in sericin are serine (30.1 g), threonine (8.5 g), aspartic acid (16.8 g) and glutamic acid (10.1 g). Fibroin is a protein mainly composed of the amino acids glycine, alanine, and serine, which form antiparallel β sheets in the spun fibres (Asakura and Kaplan, 1994; He *et al.*, 1999; Asakura *et al.*, 2002).

The structure of silk is shown in Fig. 3.1. Silk of *B. mori* is composed of the proteins fibroin and sericin, as well as soluble organic matter such as fats, wax, sand pigments and minerals. Silk is naturally coloured yellow or green and thus contains a small amount of colouring matter. Some ash will remain after silk is burned. The content of all these substances is not constant and varies within wide limits, depending on the species of silkworm and on the location and conditions of rearing. Silk filament contains the following (by total weight):

- 72–81% fibroin;
- 19–28% sericin;
- 0.8–1.0% fat and wax and
- 1.0–1.4% colouring matter and ash.

The molecular weight of the sericin ranges from 10 to 310 kDa and fibroin ranges from 300 to 450 kDa (Tanaka *et al.*, 1999; Zhou *et al.*, 2000). The key

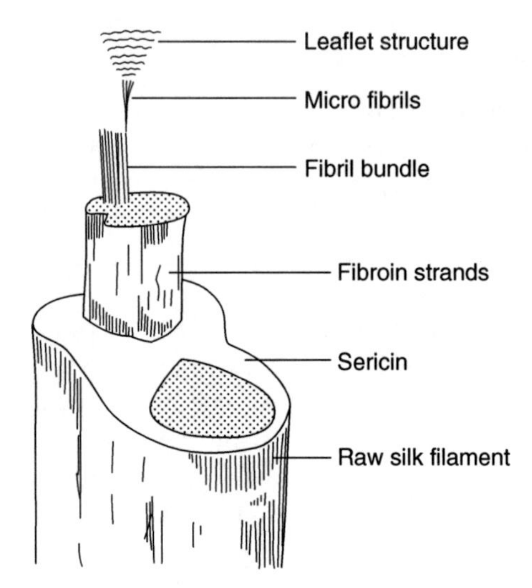

3.1 Structure of silk filament.

amino acids in sericin are serine (30.1 g), threonine (8.5 g), aspartic acid (16.8 g) and glutamic acid (10.1 g).

Fibroin is a protein and is mainly composed of the amino acids glycine, alanine and serine, which form antiparallel β sheets in the spun fibres (Asakura and Kaplan, 1994; He *et al.*, 1999; Asakura *et al.*, 2002). A small amount of cystine residue gives a very small amount of sulphur in the fibre. Fibroin has only a small number of amino acid side chains. The amounts of acid and alkali that can be absorbed by silk are relatively lower than those absorbed by wool (ca 0.2 equivalents per kg of silk). The iso-electric point of silk fibre is around pH 5. There is a small proportion of amino acid residue with the large side chains. Hydrogen bonding is important in fibroin.

3.3 Amino acid composition

The amino acid composition and sequence of amino acid residues have been found to depend on the source and the type of silk fibre (Iizuka, 1993). Amino acid composition of different varieties of silk fibroin has been studied by Lucas *et al.* (1960). Dhavalikar (1962) has also done a similar study on different varieties of Indian silks, for example mulberry, tasar, eri and muga. The findings of these two studies are tabulated in Table 3.1.

Like other fibres, silk contains crystalline and amorphous regions. There has always been an interest in finding which segments of the fibroin become part of crystalline regions. The amino acid composition of silk fibroin and wool keratin make an interesting comparison (Table 3.2). In wool keratin,

Table 3.1 Amino acid composition of different varieties of silk (mole %)

Amino acid	*B. mori* (mulberry)		*A. mylitta* (tasar)		*P. cynthia* (eri)		*A. assama* (muga)	
	1	2	1	2	1	2	1	2
Glycine	43.74	43.75	23.5	24.24	31.40	26.42	–	25.55
Alanine	28.78	29.05	36.0	39.0	47.90	35.35	–	34.34
Valine	2.16	1.85	0.80	0.67	0.57	0.54	–	0.53
Leucine	0.52	0.42	0.90	0.35	0.23	0.38	–	0.40
Isoleucine	0.65	0.53	–	0.36	0.34	0.52	–	0.39
Serine	11.88	0.00	9.80	8.82	5.10	4.96	–	7.88
Threonine	0.89	0.90	0.90	0.32	0.57	0.36	–	0.80
Aspartic acid	1.28	1.51	5.70	5.53	3.53	3.52	–	4.82
Glutamic acid	1.00	1.07	0.90	0.78	0.79	0.66	–	1.17
Phenylalanine	0.62	0.50	0.30	0.43	0.11	0.64	–	0.56
Tyrosine	5.07	5.42	4.80	4.71	5.56	5.37	–	4.61
Histidine	0.53	0.36	–	1.68	1.40	1.53	–	1.06
Arginine	1.83	1.90	13.3	11.84	1.87	6.95	–	12.25
Proline	0.35	0.50	–	0.70	0.34	0.52	–	0.47
Tryptophan	0.33	0.39	3.10	2.04	–	0.52	–	2.80
Methionine	–	–	–	–	–	0.30	–	–
Lysine	0.63	0.60	–	0.05	0.34	–	–	0.29
Cystine	–	0.08	–	0.29	–	–	–	–

1: Lucas *et al.* (1960). 2: Dhavalikar (1962).

approximately 33 mole% of the amino acids are non-polar and the rest are polar with the majority having bulky side groups. Wool is only about 25% crystalline, whereas silk is 50–60% crystalline. Consequently, the number of amino groups available for reaction in wool is much greater when compared to silk (0.82 g eq/kg in wool *vs* 0.15 g eq/kg in silk). Keratin, the principal constituent of wool, is a much more complex protein, with large side groups of all types found in appreciable quantities. Many of these contain active groups, so the side-chain linkages are important. Keratin also contains two main types of proteins: (i) low sulphur protein with relatively simple, crystallizable sections, joined to tails with a more complex chemistry containing cystine and constituting part of the non-crystalline material; (ii) high sulphur protein, which constitutes the remainder of the non-crystalline region and acts as a highly cross-linked amorphous polymer in a rubbery state. It may be observed from the results of amino acid analyses of previous studies that in *B. mori* (mulberry) silk, glycine, alanine and serine constitute about 85% of the amino acids present. However, the other amino acids, especially aromatic amino acid and tyrosine, are also found in significant amounts. It should be noted that the basic amino acids, arginine, proline, histidine and lysine, are all present in silk in small quantities.

Table 3.2 Amino acid composition of silk and wool (mole%)

Amino acid	Wool keratin	*B. mori* fibroin	*B. mori* sericin
Glycine	8.4	43.7	13.9
Alanine	5.5	28.8	5.9
Valine	5.6	2.2	2.7
Leucine	7.8	0.5	1.1
Isoleucine	3.3	0.7	0.7
Serine	11.6	11.9	33.4
Threonine	6.9	0.9	9.7
Aspartic acid	5.9	1.3	16.7
Glutamic acid	11.3	1.0	4.4
Phenylalanine	2.8	0.6	0.5
Tyrosine	3.5	5.1	2.6
Histidine	0.9	0.2	1.3
Arginine	6.4	0.5	3.1
Proline	6.8	0.5	0.6
Tryptophan	0.5	0.3	0.2
Methionine	0.4	–	0.04
Lysine	2.6	0.3	3.3
Cystine	9.8	0.2	0.1
Hydroxy amino acids	22.0	17.9	45.7
Acidic amino acids	17.2	2.3	21.1
Basic amino acids	9.9	1.0	7.7
Amino acids with sulphur containing side groups	10.2	0.2	0.1
Amino acids with polar side groups	49.1	21.2	74.6
Amino acids with non-polar side groups	33.4	76.5	24.6

In contrast to mulberry silk, the total amount of glycine, alanine and senne in wild silks is about 70%. All the non-mulberry silks are reported to have a high proportion of alanines when compared with mulberry silks. The proportion varies 36–39% in tasar, 35–45% in eri and 34–35% in muga. They have also been found to have a substantial proportion of amino acids with bulky side-groups, especially dibasic acids and arginine. According to Tsukada *et al.* (1992) the arginine content is highest in tasar (13.5%) followed by muga (12.25%). Other amino acids, *viz*, aspartic acid (5.7%), glutamic acid (0.9%) and tryptophan (3.1%), are also the highest in tasar, followed by muga. Amongst the results reported by the two studies shown in Table 3.1, notable differences may be observed in the values of glycine, alanine and arginine for eri silk. Unlike keratin, the sulphur-containing amino acids have either not been detected or are present in very small quantities. For instance, while wool has about 10–11% of such amino acids, silk fibroin seems to have around 0.1–0.2 mole% of cystine and metheonine combined. In fact, a standard chemical test used to distinguish silk from wool relates

Table 3.3 Comparative amino acid composition

Composition	*B. mori* (mulberry)	*A. pernyi* (tasar)	*A. assama* (muga)
Gly + Ala (mole%)	74.37	72.08	71.3
Gly/Ala	1.55	0.71	0.67
P/NP	0.27	0.35	0.33
LC/SC × 100	17.07	20.35	23.22

Table 3.4 Amino acid composition of silk hydrolysates (mole %)

Amino acid	Mulberry	Tasar	Eri	Muga
Glycine	40.0	7.4	12.0	14.0
Alanine	50.0	73.5	65.0	66.0
Serine	2.0	9.8	11.0	9.0
Others	15.0	10.0	11.0	17.0

Source: Nadiger *et al.* (1985).

to the absence of sulphur. It is suggested that the presence of sulphur in the form of a disulphide bond plays an important role in deciding the solubility and preventing premature formation of certain secondary structures prior to silk-fibre spinning (Gage and Manning, 1980). The polar to non-polar ratio (P/NP) of amino acid residue for muga, tasar and mulberry, along with the LC/SC (long chain/short chain) ratio, has been reported by Freddi *et al.* (1994) (Table 3.3). High P/NP ratio is an indication of the high chemical reactivity of non-mulberry silks as most of these bulky polar groups are present in amorphous regions accessible to the chemicals.

The investigation of amino acid composition in the crystalline region of silk fibroin has been carried out by Nadiger *et al.* (1985), and the results are given in Table 3.4. The crystalline regions of mulberry silk contain several repetitions of the basic sequence –(Gly–Ala–Gly–X)– where X is serine or tyrosine, while those of the non-mulberry silks are mainly of the type –(Ala) (Tsukada *et al.,* 1992). Study of the amino acid composition of the crystalline part of different silks, that is silk treated with HCl, has confirmed glycine and alanine as the major constituents in mulberry silk with a detectable amount of serine. In contrast to this result, the amino acid composition of tasar, eri and muga showed alanine to be the major constituent, with small quantities of glycine and serine (Table 3.1; Nadiger *et al.*, 1985). It appears that as yet no definite conclusion has been reached in this area.

The amino acid composition differs between varieties of silk. Three major amino acids, serine, glycine, and alanine, may be found in mulberry and non-mulberry varieties. Among the other major amino acids present are tyrosine and valine. In mulberry silks, glycine, alanine and serine

Table 3.5 Amino acid composition of silk fibres (mole %)

Amino acid	*B. mori* (mulberry)	*A. mylitta* (tasar)	*A. assama* (muga)	*Phylisamia ricini* (eri)
Aspartic acid	1.64	6.12	4.97	3.89
Glutamic acid	1.77	1.27	1.36	1.31
Serine	10.38	9.87	6.11	8.89
Glycine	43.45	27.65	28.41	29.35
Hystidine	0.13	0.78	0.72	0.75
Arginine	1.13	4.99	4.72	4.12
Threonine	0.92	0.26	0.21	0.18
Alanine	27.56	34.12	34.72	36.33
Proline	0.79	2.21	2.18	2.07
Tyrosine	5.58	6.82	5.12	5.84
Valine	2.37	1.72	1.5	1.32
Metheonine	0.19	0.28	0.32	0.34
Cystine	0.13	0.15	0.12	0.11
Isoleucine	0.75	0.61	0.51	0.45
Leucine	0.73	0.78	0.71	0.69
Phenylalanine	0.14	0.34	0.28	0.23
Tryptophan	0.73	1.26	2.18	1.68
Lysine	0.23	0.17	0.24	0.23

Source: Sen and Murugesh Babu (2004a).

together constitute about 82%, of which about 10% is serine. Tyrosine and valine are present in quantities of around 5.5% and 2.5%, respectively. The overall composition of acidic amino groups (i.e. aspartic and glutamic acids) in the mulberry variety is greater than that of the basic amino acids. Another important aspect is the composition of amino acids with bulkier side groups. The presence of bulky side groups can hamper the close packing of molecules and hinder the crystallization process. In general, a large portion of mulberry fibroin is made up of simple amino acids such as glycine and alanine, suggesting a favourable condition for crystallization (Sen and Murugesh Babu, 2003).

The total amount of glycine, alanine and serine in the non-mulberry variety is around 73%, around 10% less than in mulberry silk. All the non-mulberry silks exhibit a high proportion of alanine when compared to the mulberry variety. The proportion of alanine is about 34% in tasar, 36% in eri, and 35% in muga. These values are consistent but are lower for muga (~44%). The glycine content in these varieties is about 27–29%, which is lower than that found in the mulberry varieties (~43%).

Non-mulberry varieties also have a substantial proportion of amino acids with bulky side groups, especially aspartic acid (4–6%) and arginine (4–5%), which means that not only the acidic but also basic amino acid levels are greater. It is interesting to note the presence of sulphur-containing amino acids (i.e. cystine and methionine) in all varieties of silk. The methionine content in non-mulberry silks is slightly higher (0.28–0.34%) than that found in mulberry varieties (0.11–0.19%), whereas the cystine content is comparable (Sen and Murugesh Babu, 2003). The amino acid composition of different varieties of silk from Sen and Murugesh Babu (2004a) is shown in Table 3.5 and can be compared to the findings in Table 3.1.

3.4 Moisture regain

Mosture regain is linked to amino acid composition. The moisture regain of different varieties of silk has been reported (Sen and Murugesh Babu, 2004a). Results from different varieties of silk fibres determined under standard conditions are presented in Table 3.6. All three non-mulberry silk fibres show higher moisture regain values compared to those of the mulberry varieties. Among these, tasar shows the highest value (10.76%), followed by eri (10.21%) and muga (9.82%) for the outer layers. Mulberry bivoltine and crossbreed varieties show lower values of 8.52% and 8.63%, respectively. The higher moisture regain of non-mulberry silks suggests that all three varieties may have a higher ratio of hydrophilic to hydrophobic amino acid residues in their chemical architecture when compared to that of the mulberry varieties. It should be noted that the moisture regain of the inner layers is about 4.0–4.5% less when compared to that of the outer layers, suggesting compactness of the inner layers.

3.5 Microstructure of silk

Silk fibres (*B. mori*) spun out from silkworm cocoons consists of fibroin in the inner layer and sericin in the outer layer. Each raw silk thread has a lengthwise striation, consisting of two fibroin filaments of 10–14 μm each which are embedded in sericin. Silk fibres are biodegradable and highly crystalline with a well-aligned structure. They have a higher tensile strength than glass fibre or synthetic organic fibres, good elasticity, and excellent resilience. Silk fibre is normally stable up to 140°C and the thermal decomposition temperature is greater than 1500°C. The densities of silk fibres are in the range of 1320–1400 kg/m^3 with sericin and 1300–1380 kg/m^3 without sericin.

Scanning electron micrographs of longitudinal views of un-degummed and degummed silk fibres are presented in Figs 3.2 and 3.3, respectively. It will be observed that mulberry silk shows a more or less smooth surface (Fig. 3.3a) where as the non-mulberry silks such as tasar, muga and eri (Fig. 3.3b, 3.3c and 3.3d) all have striations on their surfaces. Scanning electron micrographs of the cross-section of silk fibres are presented in Fig. 3.4. The cross-section of silk fibre made of two types of protein (sericin and fibroin) is shown in this figure. It was found that two strands of

Table 3.6 Moisture regain (%) of silk fibres

Type of silk	Moisture regain (%)		
	Outer layer	Inner layer	% Change
Mulberry (bivoltine)	8.52	8.14	4.46
Mulberry (crossbreed)	8.63	8.28	4.05
Tasar	10.76	10.27	4.55
Muga	9.82	9.47	3.56
Eri	10.21	9.79	4.11

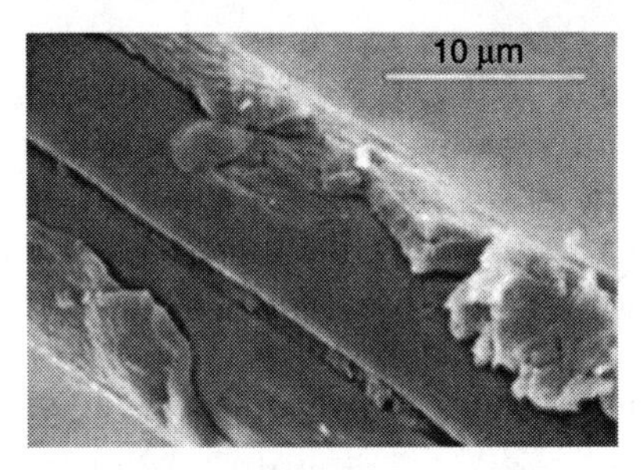

3.2 Longitudinal view of un-degummed silk fibres.

fibroin filaments were enveloped by the non-fibrous sericin. When a strand of fibroin filament is enlarged, the inner structure appears as a bundle containing a large number of fibrils (Minagawa, 2000).

There are variations depending upon the silkworm type and also among the individual cocoons. It may be seen that in this respect, mulberry and non-mulberry silks exhibit an altogether different cross-sectional morphology. The mulberry silks show a more or less triangular cross-section and a smooth surface (Fig. 3.4a). Among the non-mulberry varieties, tasar and muga exhibit an elongated rectangular or a wedge-shape cross-section and a large cross-sectional area (Fig. 3.4b and 3.4c). Eri silk has a more or less triangular shape (Fig. 3.4d). The cocoon fibres of domestic silkworms like mulberry and eri usually have an irregular cross-section ranging from triangular to circular. Even within the same fibroin filament, there may be variations in the cross section, depending upon the level of the cocoon layer.

Microstructure affects density. Sen and Murugesh Babu (2004) have reported the density values in different varieties of silks (Table 3.7). Both mulberry varieties show higher density values compared to those of non-mulberry silks. The fibre cross-sections were studied under SEM at high magnification. It was observed that mulberry varieties do not exhibit pores or voids in their cross-section and have a compact structure, whereas tasar, muga and eri silks show the presence of voids (Fig. 3.5). This could be one of

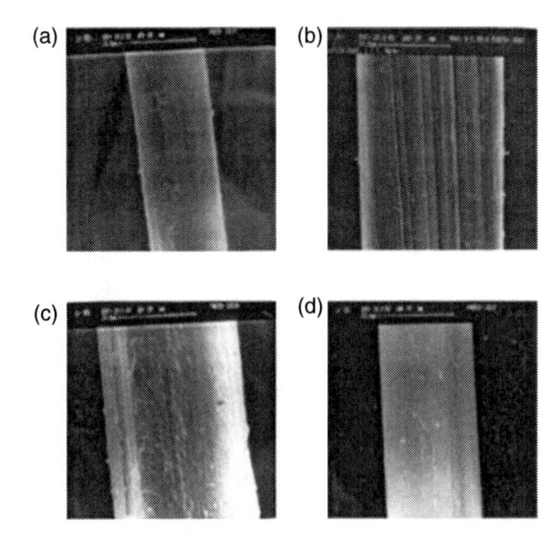

3.3 Longitudinal view of degummed silk fibres: (a) mulberry; (b) tasar; (c) muga; (d) eri.

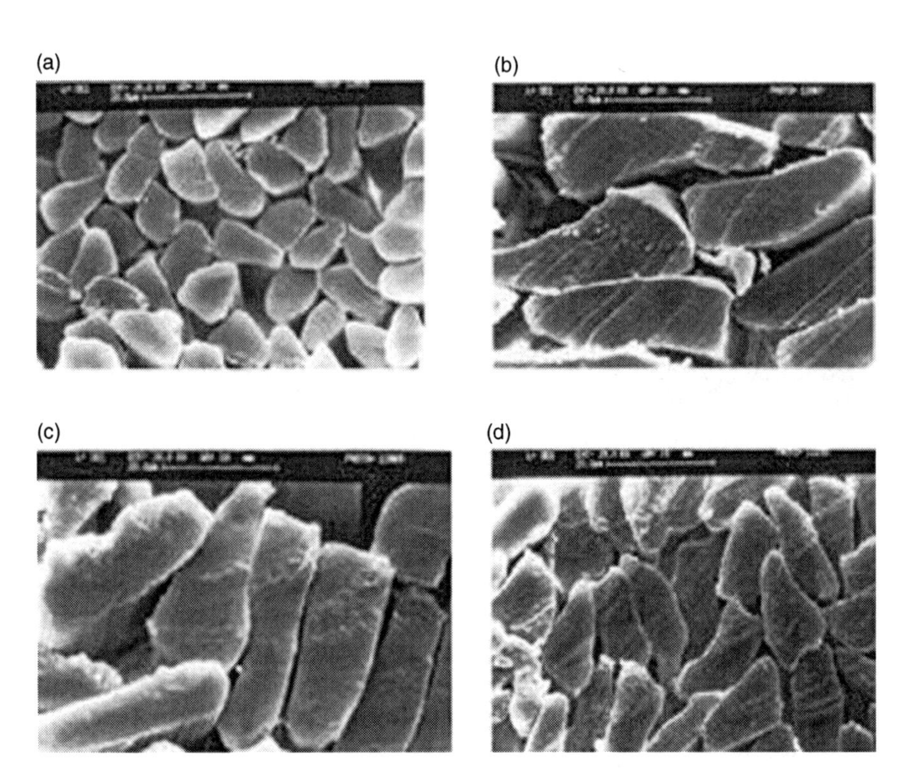

3.4 Cross-sectional view of degummed silk fibres: (a) mulberry; (b) tasar; (c) muga; (d) eri.

Table 3.7 Density ρ values for different varieties of silk

Type of silk	ρ (g/cm^3)		
	Outer layer	Middle layer	Inner layer
Mulberry (bivoltine)	1.350	1.361	1.365
Mulberry (crossbreed)	1.342	1.35	1.356
Tasar	1.300	1.33	1.340
Muga	1.332	1.34	1.348
Eri	1.28	1.29	1.295

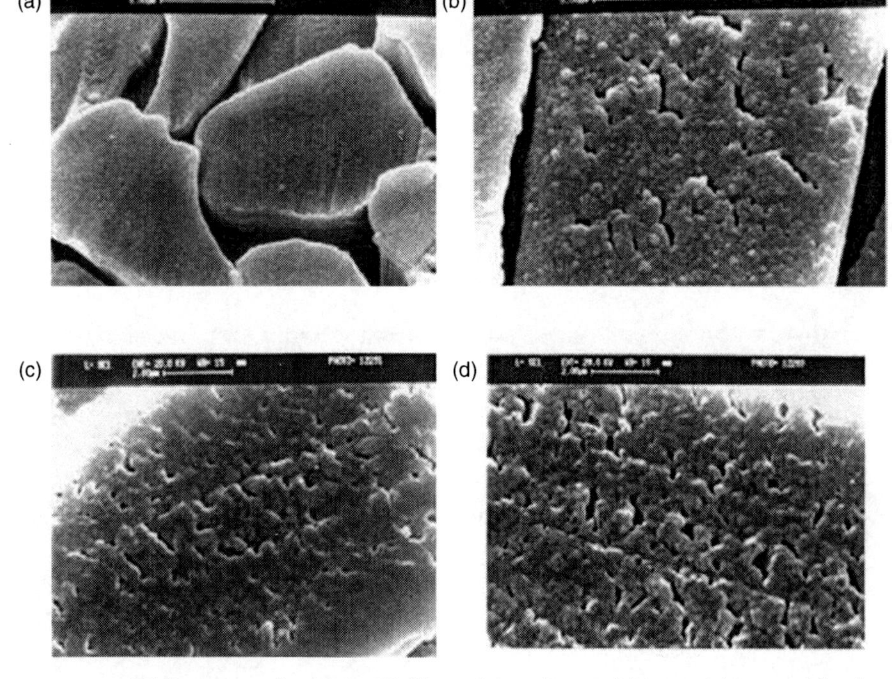

3.5 Presence of voids in silk fibres: (a) mulberry; (b) tasar; (c) muga; (d) eri.

the reasons for the higher density of mulberry varieties. The higher density values of mulberry silks also indicate a higher degree of order and compact molecular packing than those of non-mulberry silks. The mulberry bivoltine variety shows a higher density than that of the mulberry crossbreed variety. It is interesting to note that there is an increasing trend in density values from the outer to the inner layers in all the varieties. This suggests a possible increase in the degree of crystallinity and of crystallite orientation from the outer to innermost layers. Among the three non-mulberry silks, muga exhibits the highest density, followed by tasar and eri.

3.6 Chemical structure of silk

X-ray analyses of silk show the relatively high orientation of polypeptide chains along the fibre axis (Sonthisombat and Spekaman, 2004). The fibre has a two-phase system consisting of crystalline and amorphous phases. The polypeptide chain of a fibroin molecule is shown in Fig. 3.6. The crystalline phase consists of sections of polypeptide chains containing glycine, alanine and serine. These sections have simple branching owing to which the chains may be closely and compactly arranged. Sections containing residues of tyrosine, praline, diamine and dicarboxylic acids are characterized by bulky residues which impede regular and close packing of chains and, as a result, less oriented (amorphous) regions are formed. The groups found in the amorphous region are more accessible to the action of chemical reagents; for example, full saturation of the basic groups with some acids is possible without any change in the X-ray pattern of the fibre. However, the hydroxyl groups of serine residues are less accessible to the action of chemical reagents.

Molecules of silk fibroin are found in the polypeptide chain with the formula (–CHR–CO–NH–)*n*, where n = 1100 and R is one of the various amino acid residues. The polypeptide chain has both a backbone and side chains consisting of radicals R of amino acids. The degree of branching of the polypeptide chain depends on the amino acids contained in the protein. Thus the side chains form 19% of the weight in silk fibroin.

The side chains may be non-polar, as for instance hydrocarbon residues:

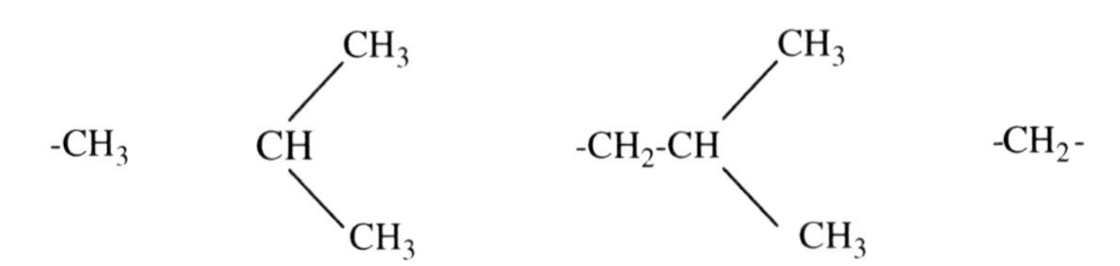

and may contain polar groups such as:

$-COOH$, $-CH_2OH$, $-C_6H_4OH$, $-SH$, $-S\text{-}S-$, NH, NH_2, $-CO\text{-}NH_2$

$$\begin{array}{c} -NH\text{-}C\text{-}NH_2 \\ \| \\ NH \end{array}$$

The physical and chemical properties of the proteins, that is the position of the isoelectric zone, their hydrolysability, swelling, solubility, etc., are largely determined by the nature of the side chains, as are the chemical reaction characteristics of the proteins. Interaction may arise between the side chains

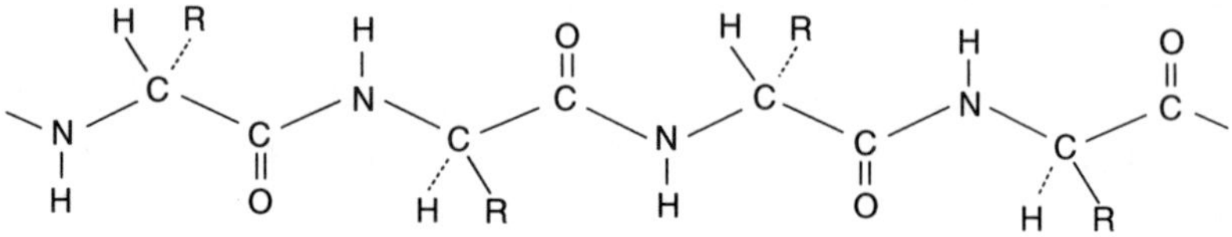

3.6 Polypeptide chain of fibroin molecule.

and different kinds of labile links, resulting in a specific spatial arrangement of the polypeptide chains or configuration of the molecule.

The primary structure is the consecutive order in which the amino acid residues are arranged in the polypeptide chains which form the protein molecule. The secondary structure is the geometrical configuration characteristic of the polypeptide chain contained in the molecular composition of the given protein, or of a separate portion of the polypeptide chain. In silk fibroin, the polypeptide chain is present in the β-form. The protein chains are stretched in contrast to the spiral α-form and the zig-zag shape of a fully extended polypeptide chain is obtained. Following detailed studies by Marsh *et al.*, (1955), the unit cell for *B. mori* silk is described as comprising the following dimensions: (a) interchain = 0.94 nm, (b) fibre axis = 0.697 nm, (c) intersheet = 0.92 nm. The unit cell of these dimensions includes four polypeptide chains or eight amino acid residues arranged anti-parallel, that is folded back and several anti-parallel chains are grouped together to form pleated sheets parallel to the a–b plane (Fig. 3.7). Each extended chain is hydrogen bonded through –C=O– and –N=H– radicals between two neighbouring chains. X-ray studies by Warwicker (1960) on a wide variety of silk proteins have shown that while the fibre axis and interchain spacing (a and b coordinates) are the same for all the fibroins, the intersheet spacings (c coordinates) vary and are related to the size of the projecting side chains.

Tertiary protein structure is the general spatial arrangement of one or several polypeptide chains (spiral, stretched or both) constituting one molecule. The tertiary structure determines the shape and size of the protein molecule. The stabilization of this structure is connected with the interaction between the side groups of the polypeptide chain. The main types of bonds for the tertiary structure are:

1. electrostatic salt bonds between positively and negatively charged groups, for instance between $–NH_3$ and –COO groups;
2. hydrogen bonds between side chains as, for instance, between hydroxyl groups and free carboxyl groups;
3. bonds due to Van der Waal's forces between polar groups and
4. bonds between non-polar side chains of valine, leucine, isoleucine, phenylalanine residues.

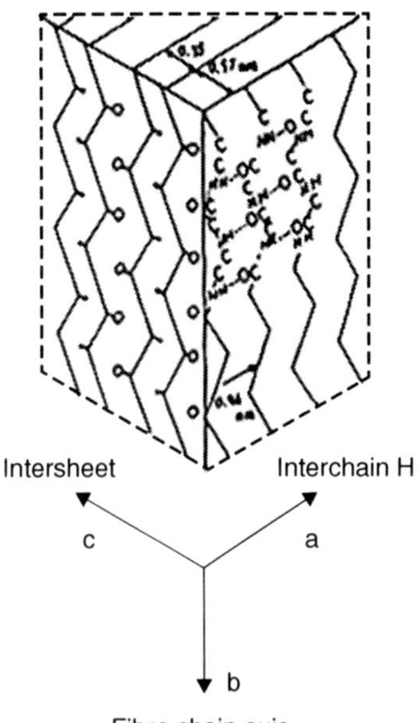

3.7 The anti-parallel β-pleated sheet configuration of *B. mori*. The chain axis is parallel to the fibre axis.

3.7 Crystalline structure of silk

Silkworm cocoon and spider dragline silks are characterized as an anti-parallel β-pleated sheet in which the polymer chain axis is parallel to the fibre axis. Other silks are known to form α-helical (those from bees, wasps and ants) or cross-β-sheet (many insects) structures. The cross-β-sheets are characterized by a polymer chain axis perpendicular to the fibre axis and a higher serine content. Most silks assume a range of different secondary structures during their processing from soluble protein in the glands to insoluble spun fibres (Kaplan, 2004). The crystalline structure of silk was first described in the 1950s as an anti-parallel, hydrogen-bonded β-sheet based on the characterization of *B. mori* fibroin (Marsh, 1955). Further modifications to this early model have been made over the years (Fraser, 1973; Colonna-Cesari, 1975).

Most silkworm cocoon and spider dragline silk fibres contain assembled anti-parallel β-pleated sheet crystalline structures (Marsh *et al.*, 1955; Lucas *et al.*, 1960; Fraser and MacRae, 1973a). Silks are considered semi-crystalline materials with 30–50% crystallinity in spider silks, 62–65% in cocoon silk fibroin from the silkworm *B. mori* and 50–63% in wild-type silkworm cocoons. In the β-sheet crystals, the polymer chain axis is parallel to the fibre axis. The polyalanine or glycine-alanine repeats are the major primary structure sequences responsible for the β-sheet formation. The β-sheets in silkworm fibre consisting of the glycine–alanine crystalline repeats are asymmetric, with one surface primarily projecting alanyl methyl groups and the other surface containing hydrogen atoms from the glycine residues (Matsumoto *et al.,* 2006).

Konishi (2000) studied the X-ray diffraction of fibroin and noticed 1 mm thick parallel silk fibroin fibres. Fibroin consists of non-crystalline and crystalline regions and the crystalline region tends to be oriented along the fibre axis because the fibre is drawn as it is extruded from the spinnerets of the silkworm. The structure of a crystalline form of *B. mori* silk fibroin which is commonly found before the spinning process (known as silk I), has been proposed as a repeated J-turn type II-like structure by combining J obtained from solid-state two dimensional spin-diffusion nuclear magnetic resonance and rotational echo double resonance (Asakura *et al.,* 2001). The molecular and crystal structure of the crystalline modifications of *B. mori*, silk I, is determined by X-ray diffraction. According to Kenji *et al.* (2001), the cell dimensions are essentially the same as those found in the synthetic model polypeptide (LAla–Gly). The molecular conformation has a 'crank-shaft' or an S-shaped zig-zag arrangement and displays remarkable agreement between the observed and calculated structure amplitudes of both dipeptide and hexapeptide sequences. It also has a reasonable level of hydrogen bond networks. X-ray structure analyses of the crystalline regions of fibroin show that the peptide chains pack in fully extended forms.

As mentioned earlier, silk is a semi-crystalline material. Generally, spider dragline and silkworm cocoon silks are considered to be semi-crystalline materials having fewer crystalline flexible chains reinforced by strong and stiff crystals (Gosline, 1986). The orb web fibres are composite materials in the sense that they are composed of crystalline regions contained within less crystalline regions estimated at 30–50% crystallinity (Iizuka, 1994; Lakshmanan and Geeta Devi, 1999). Earlier studies by X-ray diffraction analysis indicated 62–65% crystallinity in cocoon silk fibroin from silkworms, 50–63% in wild-type silkworm cocoons, and lesser amounts in spider silk (Warwicker, 1960).

Physical parameters such as crystallinity, crystallite size and crystallite orientation have been determined by X-ray diffraction. Attempts have also been made to determine the crystallinity of silk by using electron diffraction and infra-red techniques. The information derived from X-ray diffraction studies have been found useful in understanding the crystalline structure of different varieties of silks and in correlating the structure with various mechanical and thermo-mechanical properties (Bhat and Nadiger, 1980).

A study on mulberry, tasar, muga and eri (Bhat and Nadiger, 1980), showed mulberry and non-mulberry silks to exhibit different wide-angle X-ray diffraction (WAXD) patterns (Fig. 3.8). Mulberry shows a broad 2θ peak at 20°

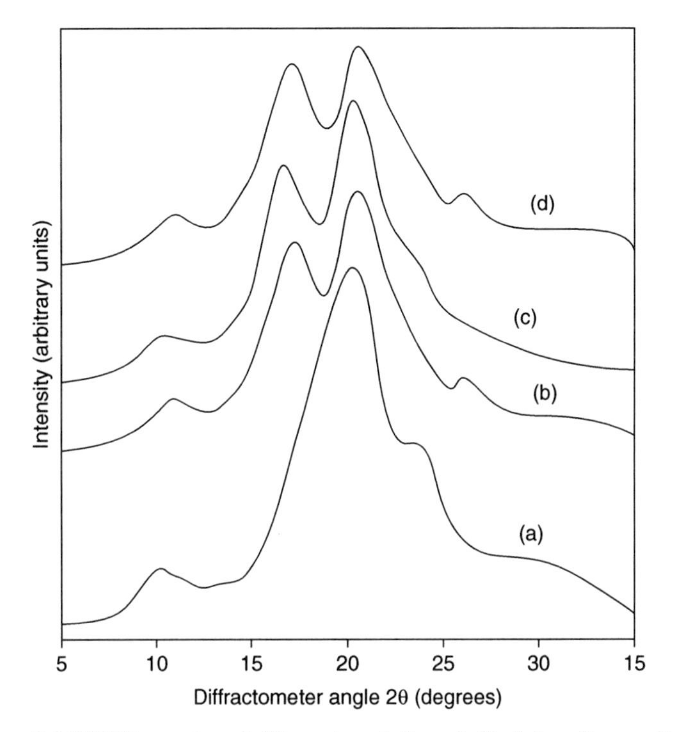

3.8 WAXD spectra of different varieties of silk: (a) mulberry; (b) tasar; (c) muga; (d) eri (Bhat and Nadiger, 1980).

corresponding to (201) reflection. All the non-mulberry silks exhibit similar X-ray diffraction patterns, showing two major peaks at 28 equal to 17.1° and 20.2° corresponding to (002) and (201) planes. Using X-ray diffraction Warwicker (1960), observed that different silk fibroins could be classified into five groups. Mulberry (*B. mori*) has been placed in group 1 and tasar, muga and eri (antheraea silks) in group 3a. Group 3a includes most antheraea species, together with other silks belonging to the family Saturniidae. The molecular conformation assumed by the fibroin chains in the crystalline regions is that of the anti-parallel p-sheet structure with unit-cell dimensions similar to that of $-(Ala)_n-$ in β-form.

Assuming the two-phase model, the method of measuring crystallinity, especially in natural fibres, has always presented difficulties. In the absence of a standard amorphous pattern, different researchers have used a variety of techniques to measure crystallinity. Manjunath *et al.* (1973) suggested a method of determining the X-ray diffraction index of crystallinity and expressed it in terms of 'lateral order'. This method was based on the fact that total order is reflected in the sharpness of diffraction peaks and the resolution 'R' of the peaks may be calculated and given by:

Table 3.8 Crystallinity indices for various silk fibres

Variety	X-ray index		IR-crystallinity index		Electron diffraction index	
	Control	Hydrolysed (96 h)	Control	Hydrolysed (96 h)	Control	Hydrolysed (96 h)
Mulberry	0.42	0.48	0.66	0.74	0.62	–
Tasar	0.43	0.77	0.50	0.66	0.60	–
Muga	0.44	0.72	0.60	0.64	0.62	–
Eri	0.43	0.65	0.50	0.82	0.63	–

$$R = \frac{2m_1}{h_1 + h_2}$$

where h_1 and h_2 are peak heights and m_1 is the height of the minimum between two peaks. If there are more than two diffraction peaks in any fibre, a generalized formula has also been given. When the resolution is completely lost, R tends to be 1, and will tend to be zero when the resolution is at a maximum. L, the lateral order has been defined as: $L = 1 - R$.

Bhat and Nadiger (1980) have reported the crystallinity and crystallite size values for silk and their hydrolysates (silks hydrolysed with HCl). The crystallinity indices determined by X-ray diffraction and electron diffraction methods have been expressed in terms of lateral order factor (Table 3.8). The crystallinity index values for un-hydrolysed samples show variations between different types of silk. It is noteworthy that silk samples treated with 6N HCl (hydrolysates) exhibited an increase in crystallinity values. The authors attribute this increase to the crystallization process caused by hydrolysis. Results obtained from different methods such as X-ray, IR-spectroscopy and electron diffraction also exhibit differences in the measured values. The data is interesting as some techniques seem to suggest the varieties are similar in this respect, while others show them to be different. The density and degree of crystallinity values in male and female varieties of mulberry, tasar and muga obtained by X-ray diffraction are reported in Table 3.9. The data have been compiled from the work of Iizuka *et al.* (1993a, 1993b, 1993c, 1994, 1996, 1997). No definite trend may be observed between the male and female varieties and no significant differences exist between the mulberry and the non-mulberry varieties. However, the data obtained for the temperate tasar, muga and mulberry (extra fine) may be compared with those for un-hydrolysed samples obtained by Bhat *et al.* (1980). No available literature explains the differences in crystallinity or lateral order along the length of the silk filament. All the values appear to be average values. As some studies have reported a gradual reduction

Table 3.9 Density and crystallinity values for various silk fibres

Variety	Sex	Density (g/cm^3)	Percentage crystallinity (X-ray)
A. myllita	M	1.329	39.5
(tropical tasar)	F	1.334	39.6
A. proylei	M	1.345	44.6
(temp.tasar)	F	1.344	41.8
A. assama	M	1.314	42.7
(muga)	F	1.327	43.7
Ariake	M	1.354	37.2
(mulberry extra coarse)	F	1.356	35.2
Shunreix shongetsu	M	1.333	43.1
(mulberry extra fine)	F	1.320	41.9

Table 3.10 Average crystallite size for silk and their hydrolysates

Variety of silk	Planes	Control sample (Å)	48-h hydrofibroin (Å)	(48 + 48) h hydrofibroin (Å)
Mulberry	002	10	15	20
	201	19	20	25
Tasar	002	27	68	60
	201	47	56	52
Eri	002	30	56	56
	201	47	47	60
Muga	002	32	47	59
	201	39	43	47

of denier, elongation-at-break, degumming loss, etc., in the inner layers of cocoons, it is reasonable to expect the change in the above parameters to occur along the length of the filaments.

Bhat *et al.* (1980) investigated the crystallite size of hydrolysates (silk treated with HCl for different durations) of Indian silk varieties by means of wide angle X-ray diffraction (Table 3.10). The results show that non-mulberry silks appear to have larger crystals than mulberry silk. This may be due to the higher alanine content and the predominance of Ala–Ala links in the crystalline region. The researchers concluded that following HCl treatment, the crystallite size calculated from 002 plane increased from 27 Å to 60 Å in tasar, 30 Å to 56 Å in eri and 32 Å to 59 Å in muga. It is interesting to note that the crystallite size in mulberry control fibres was 10 Å from 002 plane, and the size increased to 20 Å after a 96-h treatment. There was no significant increase in the crystallite size of mulberry silk calculated from the 002 plane, whereas all three non-mulberry silks showed a considerable increase in their crystallite size for 002 plane after the acid treatment. Similar trends were also observed from the data obtained for the 201 plane. The authors' opinion is that in the case of wild silks, the process of acid hydrolysis causes the Ala–Ala segments to align parallel to the existing micro-crystal, thus causing the crystal to grow. However, the greater crystallite size may also be related to the different crystal geometry of non-mulberry and mulberry silks and requires further investigation.

In an extension to the above study, Somashekarappa *et al.* (1996) conducted a study on the crystal size distributions of physically and chemically modified *B. mori* and tasar silk fibres. The first part of the study consisted of the mechanical stretching of wet degummed mulberry and tasar yarns to differing levels. In the second, the mulberry silk yarn was treated with 5–20% (w/v) formic acid and zinc chloride for 1 h at boil. It was observed that when *B. mori* silk was stretched by around 19% and dried under tension, the crystal size marginally increased from a value of 10.69 Å to 11.47 Å. There was a significant increase in the crystal size values of tasar silk yarn which increased to 15.03 Å (treated yarn) from 8.82 Å (untreated yarn). *B. mori* silk yarn treated with 20% formic acid showed a large increase in crystal size, to a value of 34.01 Å when compared to those treated with 20% zinc chloride solution, resulting in a value of 23.90 Å. The authors concluded that the physical and chemical treatment of silk fibres results in a substantial increase in crystal size values.

Although silk may be categorized as a non-thermoplastic fibre, some studies on the effect of heat treatment show interesting results. In a study on the crystal size parameters of annealed bivoltine mulberry silk fibres (Somashekar and Gopalakrishna Urs, 1995), the un-degummed silk fibres were annealed at 100°C, 140°C and 200°C in air for differing lengths of time and without stretching the fibre. By using Fourier analysis of the scattered X-ray reflections, the authors determined the crystal size lattice distortion parameters for (201) equatorial reflections. The authors' conclusion is that the intensity profile of the X-ray reflection of a semi-crystalline sample such as natural silk is a function of the crystal size distribution and the lattice distortion. The intensity $I(s)$ is given by the equation,

$$I(s) = \sum A(n)\cos\{2\pi nd(s - s_0)\}$$

where s_0 is the value of s (= sine θ/λ) at the peak of the profile; d is the mean spacing of the lattice planes causing the reflection; and n is the harmonic number. The Fourier coefficient $A(n)$ can be factorized to size $A_s(n)$ and disorder coefficient $A_d(n)$:

$$A(n) = A_s(n) \cdot A_d(n)$$

By taking the exponential distribution function for crystal size, the previous equation may be written as:

$$A_s(n) = A(0)\left[\frac{\exp\{-\alpha(n-p)\}}{\alpha<N>}\right]$$

where $<N>$ is the average number of unit cells in a column with the crystal direction normal to the lattice planes causing reflection. Here, p is the smallest number of unit cells in a column and, α is a parameter defining the width of the exponential distribution function of column lengths and is given by:

$$\alpha = \frac{1}{<N>-p}$$

The average crystal size $<D>$ is given by,

$$<D> = <N>d_{\mathrm{hkl}}$$

The results of the above study showed that the crystal size for bivoltine mulberry silk fibres annealed at 100°C for seven hours increased from 24 Å to 25 Å. However, there was no significant change in the crystal size above 100°C. It was concluded that the annealing of silk fibres under these conditions improves the crystal size of silk fibres without affecting the lustre. Although the authors have given the problem extensive mathematical treatment, the data probably loses much of its significance as the work appears to have been done on un-degummed fibres. The crystallization tendencies of fibroin and sericin are visibly different and will therefore give confusing and distorted results.

In another study, Tsukada *et al.* (1992) reported the effect of thermal treatment on crystalline structure and orientation, using WAXD spectrum. They reported that the crystalline structure of *B. mori* fibres remained unchanged, regardless of the heat treatment. However, in the case of tasar, the X-ray diffraction maximum exhibited a slight shift to smaller angles above 230°C. The authors concluded that the crystalline structure of *B. mori* and tasar silk fibres was not significantly affected by thermal treatment and remained essentially unchanged. It appears from the literature that the studies on crystallinity and crystallite size have been mostly confined to *B. mori* (mulberry) and tasar silks. Very few studies have been made on other non-mulberry silks such as muga and eri.

3.7.1 Studies of crystalline structure using infra-red (IR) spectroscopy

Infra-red studies on silk fibroin (Magoshi *et al.*, 1979; Bhat and Ahirrao, 1983, 1985; Nadiger and Bhat, 1985; Baruah *et al.*, 1991) suggest that the appearance of absorption bands at 1660, 1540, 1235 and 650 cm^{-1} correspond to amide 1, amide 2, amide 3 and amide 4, respectively. The characteristics of the random coil conformation (amorphous) and those appearing at 1630, 1535, 1265 and 700 cm^{-1} indicate p-conformation (crystalline).

The calculation of the crystallinity index by IR-spectroscopy has been carried out by using the ratio of intensity bands corresponding to the crystalline region with those of the amorphous region. Drukker *et al.* (1953) have reported a crystallinity value of 63% for *B. mori* silk using bands at 1528 and 1560 cm^{-1}. However, the authors emphasized that the value obtained was not exact as the resolution of these bands was not particularly good. This value appears high when compared to those obtained from WAXD. Similar studies on mulberry and wild varieties of Indian silk were conducted (Bhat *et al.*, 1980) using the ratio of absorbance of bands at 1265 cm^{-1} (β-form) and 1235 cm^{-1} [α- (random coil conformation) form]. It may be observed from the data (Table 3.8) that mulberry and muga silks showed higher crystallinity indices (0.66 and 0.60 respectively) when compared to tasar (0.50) and eri (0.50).

In a study on the crystallinity of silk hydrolysates, Bhat *et al.* (1980) reported the IR spectra of different varieties of silk in the range of 1400–800 cm^{-1} (Fig. 3.9). They assigned the band at 1015 cm^{-1} to gly–gly linkage, the band at 970 cm^{-1} to ala–ala linkage and those at 998 and 975 cm^{-1} to ala–gly linkages. In mulberry, the bands at 1015 and 970 cm^{-1} were entirely absent, while the bands at 975 and 998 cm^{-1} were present. The non-mulberry silks showed a strong band at 970 cm^{-1}. The intensity of these bands at respective wave numbers was further enhanced in hydrolysed silk samples of both mulberry and non-mulberry varieties. The band at 1015 cm^{-1}, which corresponded to gly–gly linkage, totally disappeared in the hydrolysed samples of wild silk varieties. On the basis of the above results, it was concluded that the crystalline structure of mulberry must have been largely formed by the sequence of glycine and alanine linkages, and that of wild silks by nonlabile ala–ala linkages.

Studies on mulberry, muga and eri (Baruah *et al.*, 1991) further supported the above conclusions by observing predominant absorption bands at 980 and 998 cm^{-1} for mulberry. Muga and eri showed strong bands at 970 cm^{-1}. α and β-forms of molecular conformations were also observed in the above fibres. Based on these observations, it was concluded that the absorption band at 980 cm^{-1} for mulberry represents the Ala–Aly linkage, which predominates in the crystalline region of the fibre. The band at 970 cm^{-1}, for

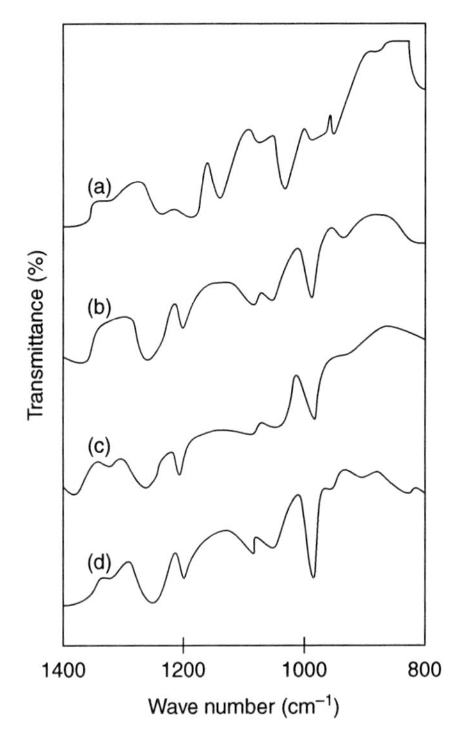

3.9 Infra-red (IR) spectra of different varieties of silk: (a) mulberry; (b) tasar; (c) muga; (d) eri.

both muga and eri, represents the Ala–Ala linkage indicating a crystalline phase similar to poly-L-alanine linkage observed by previous workers. In light of the above, it is difficult to deduce which type of silk has the higher crystallinity. Similarly, no conclusion can be drawn as to which technique is more accurate, as the trends also appear to change. However, the data do point to the fact that mulberry varieties have predominantly glycine–alanine linkages in the crystalline region, while non-mulberry varieties demonstrate the predominance of alanine–alanine linkages.

3.8 Optical properties of silk

Silk fibroin extracted from silkworm cocoons is a unique bio-polymer combining biocompatibility and implantability with excellent optical properties. Silk may be used as an optical material in applications for biomedical engineering, photonics and nano-photonics. It can be nano-patterned with features smaller than 20 nm which allows the manufacture of structures including holographic gratings, phase masks, beam diffusers and photonic crystals out of a pure protein film. The properties of silk allow these devices

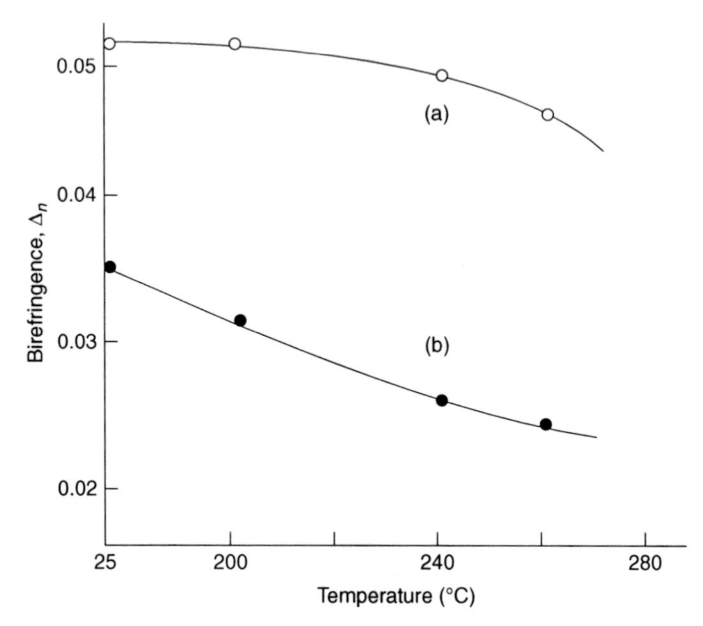

3.10 Effect of treatment temperature on birefringence of silk fibres: (a) *B. mori;* (b) *Antheraea* silk.

to be 'biologically activated', so offering new opportunities for sensing and bio-photonic components. Many interesting bio-optical devices may be fabricated by doping silk films with fluorescent materials (such as quantum dots shown above). Means of enhancing the light emission by patterning silk film surfaces, making tunable wavelength devices and printing specific patterns on silk film surfaces may also be explored.

The lustre associated with silk is partly due to the influence on the pattern of light reflected from its triangular shape. In an attempt to understand the optical properties of silk, many researchers have determined the refractive index and birefringence of fibres. The refractive index of silk usually varies throughout its cross-section. The birefringence (η) values vary between 0.051 and 0.0539 in mulberry silk and between 0.030 and 0.034 in non-mulberry silks (Tsukada *et al.*, 1994).

In a study on the structural changes induced by heat treatment, Tsukada *et al.* (1992) investigated the effect of temperature on the crystalline structure, birefringence and isotropic refractive index of *B. mori* and *Antheraea pernyi* (tasar) silk fibres. It was found that the birefringence, which is regarded as a measure of the average orientation of the molecules in a fibre, did not change noticeably in *B. mori* fibres, regardless of heat treatment in the range from 25°C to 240°C. As the temperature exceeded 240°C, the Δ_n value showed a slight decrease from 0.053 - 0.048 (Fig. 3.10).

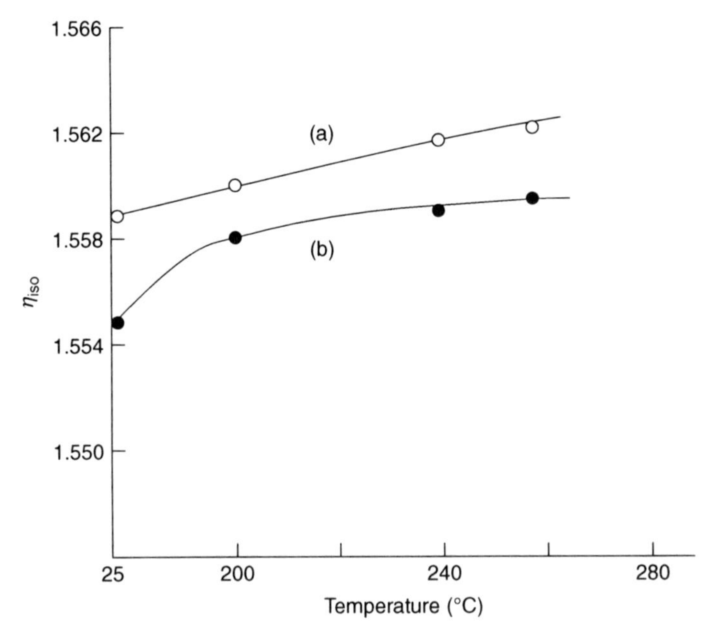

3.11 Effect of temperature on isotropic refractive index, η_{iso}, of silk fibres: (a) *B. mori;* (b) *Antheraea* silk.

However, it is interesting to note that tasar silk fibres exhibited a markedly lower birefringence value of 0.035–0.025, suggesting a poor structural stability under the influence of heat when compared to *B. mori* fibres. The degree of orientation of tasar silk was found to decrease linearly with temperature.

The isotropic refractive index is another important parameter. It is related to the crystallinity of fibres (η_{iso}) and, when calculated by the Beckeline method, exhibited a linear increase for *B. mori* from 1.559 to 1.563 in the temperature range 25–240°C (Fig. 3.11). The rate and extent of increase of niso in tasar was quite different from that of *B. mori.* Up to 200°C, the value increased markedly from 1.555 to 1.558 and became almost constant above 200°C. It was concluded that the crystalline structure of both mulberry and tasar silks did not show any significant changes after heat treatment. The heat treatment had no significant effect on the birefringence of mulberry fibres, while tasar silk exhibited a markedly lower birefringence, implying a lower molecular orientation in the amorphous region but not in the crystalline region. The findings suggested that *B. mori* silk possesses a high degree of order and molecular orientation in the amorphous region when compared to tasar fibres. The results of these studies are presented in Table 3.11. Das (1996) also showed that the orientation index in mulberry silks is higher.

Table 3.11 Average crystallite size for silk and their hydrolysates

	Birefringence (Δ_n)		η_{iso}	Orientation index (%)
Type of silk	1	2	1	2
Mulberry	0.051	0.0539	1.559	85
Tasar	0.034	0.0165	1.542	61
Muga	0.030	–	1.557	–

1: Tsukada *et al.* (1992). 2: Das (1996).

In a study on the comparative structural characterization of natural and laboratory-produced silk fibres of *B. mori* fibroin, Trabbic and Yager (1998) reported the crystallite orientation of the above two silk fibres. The degree of preferential crystal orientation was determined from the fibre X-ray diffraction patterns. The azimuthal width at half the maximum intensity of the strongest equatorial reflection was used as an empirical measure of crystalline orientation. The authors referred to this as parallelism (Π), which was calculated as follows:

$$\Pi = \left[1 - \left(\frac{H^\circ}{180^\circ}\right)\right] \times 100$$

where H° is the azimuthal angular width in degrees at half-maximum intensity. Π = 0 signifies a completely unoriented sample and Π = 100 signifies perfectly aligned, parallel crystals. The results of this study indicate that crystallite orientation increased with an increase in the draw ratio. Similar conditions were reproduced in laboratory-produced silk at a draw ratio of 2.5 and no further increase was observed above this value. It appears from the literature that no work on the above aspects has been undertaken for muga and eri.

The birefringence of different varieties of silk fibres has been reported by Sen and Murugesh Babu (2004), using the Beckeline method. A series of mixtures using two liquids (i.e. liquid paraffin (refractive index, η = 1.465) and 1-chloronaphthaline (η = 1.633)) were prepared for determination of the refractive index of single filaments in both the parallel and the perpendicular directions. The refractive index of the mixture, in which the Beckeline vanished under a polarizing microscope, was taken to be the refractive index of fibre. The birefringence (Δ_n) was calculated using the equation:

$$\Delta_n = \eta_{\parallel} - \eta_{\perp}$$

Table 3.12 Birefringence (Δ_n) and sonic modulus (S_ϵ) values for different varieties of silks

	Δ_n			S_ϵ (g/d)		
Type of silk	Outer layer	Middle layer	Inner layer	Outer layer	Middle layer	Inner layer
Mulberry (bivoltine)	0.054	0.055	0.056	165	170	206
Mulberry (crossbreed)	0.051	0.052	0.052	153	167	192
Tasar	0.041	0.042	0.042	106	107	115
Muga	0.040	0.041	0.042	103	108	114
Eri	0.034	0.035	0.035	101	106	109

where $\eta_\parallel$ is the refractive index of the fibre in the parallel direction and $\eta_\perp$ is the refractive index of the fibre in the perpendicular direction to the plane of the polarizer light. An average of five readings was reported.

The birefringence (Δ_n) value gives an idea of the overall molecular orientation and the sonic modulus (S_c) gives the combined effect of order and orientation. The values of Δ_n and S_c are listed in Table 3.12. Both birefringence and sonic modulus were found to increase from the outer to the inner layers within the same variety. This trend was observed in all varieties.

3.9 References

Altman, G.H., Diaz, F., Jakuba, C., Calabro, T., Horan, R.L., Chen, J., Lu, H., Richmond, J. and Kaplan, D.L. (2003), Silk-based biomaterials, *Biomater.*, **24**, 401–416.

Asakura, T. and Kaplan, D.L. (1994), Silk production and processing, in *Encyclopedia of Agricultural Science*, ed. Arntzen, C.J. and Ritter, E.M., Academic Press, New York, **4**, 1–11.

Asakura, T., Yamare, T., Nakazawa, Y., Kamada, T. and Ando, K. (2001), Structure of *Bombyx mori* silk fibroin before spinning in solid state studied with wide angle X-ray scattering and 13 cross polarization/magic angle spinning NMR, *Biopolym.*, **58**, 521–525.

Asakura, T., Yao, J., Yamane, T., Umemura, K. and Ulrich, A.S. (2002), Heterogeneous structure of silk fibers from *Bombyx mori* resolved by 13C solid-state NMR spectroscopy, *J. Am. Chem. Soc.*, **124**, 8794–8795.

Baruah, G.C., Talukdar, C. and Bora, M.N. (1991), Infrared spectroscopic study of some natural silk fibres, *Ind. J. Phys.*, **65B**(6), 651–654.

Becker, N., Oroudjev, E., Mutz, S., Cleveland, J.P., Hansma, P.K., Hayashi, C.Y., Makarov, D.E. and Hansma, H.G., (2003), Molecular nanosprings in spider capture-silk threads, *Nat. Mater.*, **2**, 278–283.

Bell, F.I., McEwen, I.J. and Viney, C. (2002), Fibre science: supercontraction stress in wet spider Dragline, *Nature*, **416**, 37.

Bhat, N.V. and Ahirrao, S.M. (1983), *J. Appl. Polym. Sci.*, **28**, 1273–1280.

Bhat, N.V. and Ahirrao, S.M. (1985), *Text. Res. J.*, **55**(1), 65–71.

Bhat, N.V. and Nadiger, G.S. (1980), *J. Appl. Polym. Sci.*, **25**, 921–932.

Carboni, P. (1952), *Silk, Biology, Chemistry and Technology*, Chapman & Hall, London.

Colonna-Cesari, F., Premilat, S. and Lotz, B. (1975), *J. Mol. Biol*, **95**, 71.

Craig, C.L. (1997), Evolution of arthropod silks, *Ann. Rev. Entomol.*, **42**, 231–67.

Craig, C.L., Hsu, M., Kaplan, D. and Pierce, N.E. (1999), A comparison of the composition of silk proteins produced by spiders and insects, *Int. J. Biol. Macromol.*, **24**, 109–181.

Dhavalikar, R.R. (1962), *J. Sci. Ind. Res.*, **21C**, 261.

Drukker, B., Hainsworth, R. and Smith, S.G. (1953), *J. Text. Inst.*, **44**, T420.

Fraser, R.D.B. and MacRae, T.P. (1973a), *Conformation in Fibrous Proteins and Related Synthetic Polypeptides*, Academic Press, New York.

Fraser, R.D.B. and MacRae, T.P. (1973b), Silks. In *Conformation in Fibrous Proteins and Related Synthetic Polypeptides*, Chapter 13, Academic Press, New York.

Freddi, G., Gotoh, Y., Mori, T., Tsutsui, I. and Tsukada, M. (1994), *J. Appl. Polym. Sci.*, **52**, 775–781.

Gage, L.P. and Manning, R.F. (1980a), *J. Biol. Chem.*, **255**, 9444.

Gage, L.P. and Manning, R.F. (1980b), *J. Biol. Chem.*, **255**, 9451.

Gosline, J.M., DeMont, M.E. and Denny, M.W. (1986), *Endeavour*, **10**, 37.

He, S.J., Valluzzi, R. and Gido, S.P. (1999), Silk I structure in *Bombyx mori* silk foams, *Int. J. Biol. Macromol.*, **24**, 187–195.

Iizuka, E. (1993), Spinning process and physical properties of silk thread. In *Chemical Processing of Silk*, ed. Gulrajani, M.L., IIT, Delhi.

Iizuka, E. (1994), *Int. J. Wild Silkmoth Silk*, **1**(2), 143–146.

Iizuka, E. and Itoh, H. (1997), *Int. J. Wild Silkmoth Silk*, **3**, 37–42.

Iizuka, E., Kawano, R., Kitani, Y., Okachi, Y., Shimizu, M. and Fukuda, A. (1993a), *Ind. J. Seric.*, **32**(2), 27–36.

Iizuka, E., Okachi, Y., Ohabayashi, S., Fukuda, A. and Shimizu, M. (1993b), *Ind. J. Seric.*, **1**(1), 1–8.

Iizuka, E., Okachi, Y., Shimizu, M., Fukuda, A. and Hashizume, M. (1993c), *Ind. J. Seric.*, **32**(2), 175–183.

Iizuka, E, Sekiguchi, S., Okachi, Y. and Ohbayashi, S. (1996a), *Int. J. Wild Silkmoth Silk*, **2**, 5–10.

Iizuka, E., Teramoto, A., Lu, Q., Si-sia Min and Shimizu, O. (1996b), *J. Seric. Sci. Jpn.*, **65**(2), 134–136.

Kaplan, D.L. (2004), 'Silk' in *A Public Scientific and Technical Research Establishment (EPST)*, **11**, 841–850.

Kenji, O., Somashekar, R., Noguchi, K. and Syuji, I. (2001), Refined molecular and crystal structure of silk I based on ala–gly and (ala–gly)2 ser–gly peptide sequence, *Biopolymers*, **59**, 310–319.

Konishi, T. (2000), Structure of fibroin – 8. In *Structure of Silk Yarn*, ed. Hojo, N., Oxford and IBH publication Co. Pvt. Ltd., New Delhi, 267–277.

Lakshmanan, S. and Geeta Devi, R.G. (1999), *Ind. Tex. J.*, **6**, 22–24.

Lucas, F., Shaw, J.T.B. and Smith, S.G. (1960), Comparative studies of fibroins. I. The amino acid composition of various fibroins and its significance in relation to their crystal structure and taxonomy, *J. Mol. Biol.*, **2**, 339–349.

Magoshi, J., Mizuide, M. and Magoshi, Y. (1979), *J. Polym. Sci.*, **17**, 515–520.

Manjunath, B.R., Venkataraman, A. and Stephen, T. (1973), *J. Appl. Polym. Sci.*, **17**, 1091–1099.

Marsh, R.E., Corey, R.B. and Pauling, L. (1955a), *Biochim. Biophys. Acta*, **16**, 1.

Marsh, R.E., Corey, R.B. and Pauling, L. (1955b), An investigation of the structure of silk fibroin, *Biochim. Biophys. Acta*, **16**, 1–34.
Matsumoto, A., Kim, H.J., Tsai, I.Y., Wang, X., Cebe, P. and Kaplan, D.L. (2006), Silk. In *Hand Book of Fibre Chemistry*, ed. Lewin, M., Taylor & Francis Group, New York.
Minagawa, M. (2000), Science of silk. In *Structure of Silk Yarn*, Vol. I, ed. Hojo, N., Oxford & IBH Publishing Co. Pvt. Ltd, New Delhi, 185–208.
Nadiger, G.S. and Bhat, N.V. (1985), *J. Appl. Polym. Sci.*, **30**, 4127–4136.
Sen, K. and Murugesh Babu, K. (2004a), Studies on Indian silk. I. Macrocharacterization and analysis of amino acid composition, *J. Appl. Polym. Sci.*, **92**, 1080–1097.
Sen, K. and Murugesh Babu, K. (2004b), Studies on Indian Silk. II. Structure–property correlations, *J. Appl. Polym. Sci.*, **92**, 1098–1115.
Sheu, H.S., Phyu, K.W., Jean, Y.C., Chiang, Y.P., Tso, I.M., Wu, H.C., Yang, J.C. and Ferng, S.L. (2004), Lattice deformation and thermal stability of crystals in spider silk, *Int. J. Biol. Macromol.*, **34**, 325–331.
Somashekar, R. and Gopalakrishna, U. (1995), *Polymer*, **36**(10), 2007–2011.
Somashekarappa, H., Selvakumar, N., Subramanium, V. and Somashekar, R. (1996), *J. Appl. Polym. Sci.*, **59**, 1677–1681.
Sonthisombat, A. and Spekaman, P.T. (2004), *Silk: Queen of Fibres – The Concise Story*, Department of Textile Engineering, Faculty of Engineering, Rajamangala University of Technology Thanyaburi (RMUTT).
Tanaka, K., Kajiyama, N., Ishikura, K., Waga, S., Kikuchi, A., Ohtomo, K., Takagi, T. and Mizuno, S. (1999), Protein Structure M, *Biochem. Biophys. Acta*, **1432**, 92–103.
Trabbic, K.A. and Yager, P. (1998), Comparative structural characterization of naturally- and synthetically-spun fibers of *Bombyx mori* fibroin, *Macromolecules*, **31**(2), 462–471.
Tsukada, M., Freddi, G., Nagura, M., Ishikawa, H. and Kasai, N. (1992), *J. Appl. Poly. Sci.*, **46**, 1945–1953.
Warwicker, J.O. (1960), *J. Mol. Biol.*, **2**, 350.
Yao, J. and Asakura, T. (2004), Silks, *Encyclopedia of Biomaterials and Biomedical Engineering*, Marcel Dekker, New York, 1363–1370.
Zhou, C.Z., Confalonieri, F., Medina, N., Zivanovic, Y., Esnault, C., Yang, T., Jacquet, M., Janin, J., Duguet, M., Perasso, R. and Li, Z.G. (2000), *Nucleic Acids Research*, **28**, 2413–2419.

4
Mechanical and thermal properties of silk

DOI: 10.1533/9781782421580.84

Abstract: This chapter reviews the mechanical and thermal properties of silk, including tensile properties and visco-elastic behaviour.

Key words: mechanical properties of silk, thermal properties of silk, tensile properties, visco-elastic behaviour.

4.1 Introduction

Silkworms produce cocoons from silk-protein-based fibres as a means of protection during their metamorphosis into moths, and web-weaving spiders produce a number of different silk-protein-based fibres to capture prey (in webs), to protect/preserve their offspring/prey (in cocoons) and as lifelines to escape from predators. Certain silk fibres have mechanical properties superior to nylon, Kevlar and high-tensile steel. Naturally occurring silkworm and spider silk fibres have been used by humans for millennia for applications as diverse as currency, hunting (bow strings, fishing lines or nets), paper, textiles and wound dressings owing to their mechanical properties and biocompatibility (Scheibel, 2004, 2005; Hardy, *et al*, 2008; Table 4.1).

Silk fibres are remarkable materials displaying unusual mechanical properties: strong, extensible and mechanically compressible. They also display interesting thermal and electromagnetic responses, particularly in the UV range, and form crystalline phases related to processing. The mechanical properties of silk fibres are a direct result of the size and orientation of the crystalline domains, the connectivity of these domains to the less crystalline domains, and the interfaces or transitions between less organized and crystalline domains. Other properties of silk such as good thermal stability, optical responses, dynamic mechanical behaviour and time-dependent responses have all been used in number of applications in various fields. In this chapter, details of various properties of different varieties of silk are presented.

Table 4.1 Mechanical properties of natural silk and man-made fibres

Material	Tensile strength (MPa)	Elongation-at-break (%)	Toughness (MJ/m^3)
Bombyx mori cocoon silk	600	18	70
Araneus diadematus lifeline silk	1100	27	160
Nylon	900	18	80
Kevlar 49TM	3600	2.7	50
High-tensile steel	1500	0.8	6

4.2 Tensile properties

The tensile properties of different varieties of silks in terms of tenacity, elongation-at-break and initial modulus have been determined by a number of workers (Iizuka *et al.*, 1993a; Freddi *et al.*, 1994; Iizuka, 1994, 1995; Iizuka and Itoh, 1997). Studies conducted on some mulberry and non-mulberry varieties by Iizuka *et al.*, reveal that the tenacity, elongation and modulus are all dependent on the linear density of the filament, and the linear density or the mean size in turn depends on the silkworm race. The tenacity is found to be linearly related to the linear density of the filament (Fig. 4.1). The correlation is negative, i.e. as the linear density increases, tenacity decreases. A similar trend has also been observed for modulus. Elongation, on the other hand, increases with an increase in linear density. The tenacity ranges between 2.5 and 4.82 g/d (grams per dernier), for Japanese and Chinese mulberry varieties, 2.4–4.32 g/d for Indian mulberry varieties, 3.74–4.6 g/d for Indian tasar varieties (Iizuka, 1995). In a study on chemical structure and physical properties of *Antheraea assama* (muga) silk, it has been reported that the tenacity of muga varies between 3.2 and 4.95 g/d (Iizuka *et al.*, 1993b; Freddi *et al.*, 1994). Another important non-mulberry variety, eri, showed lowest tenacity value, ranging between 2.3 and 4.0 g/d (Iizuka and Itoh, 1997).

Elongation-at-break, on the other hand, showed a higher value for all the non-mulberry silks compared to mulberry varieties. The values range between 31% and 35% for tasar, 34% and 35% for muga and 29% and 34% for eri silks respectively. The elongation values for mulberry varieties range between 19% and 24%. Some of the mechanical properties of different varieties of silk are summarized in Table 4.2.

Lucas *et al.* (1955) attempted to correlate the tensile behaviour of different silk fibroins with the ratio (100 × LC/SC) of total short side-chain (SC) amino acids to the total long side-chain amino acids (LC) present in the fibroin. The amino acids, glycine, alanine, serine and threonine are considered as SC amino acids and remaining bulky side group amino acids are added to get LC amino acids. Accordingly, mulberry, having a ratio of 13.5, has a low elongation to break than tasar in which the ratio is 22.1–30.0. Table 4.3 summarizes the data on the ratio of LC/SC × 100, tenacity and elongation of different

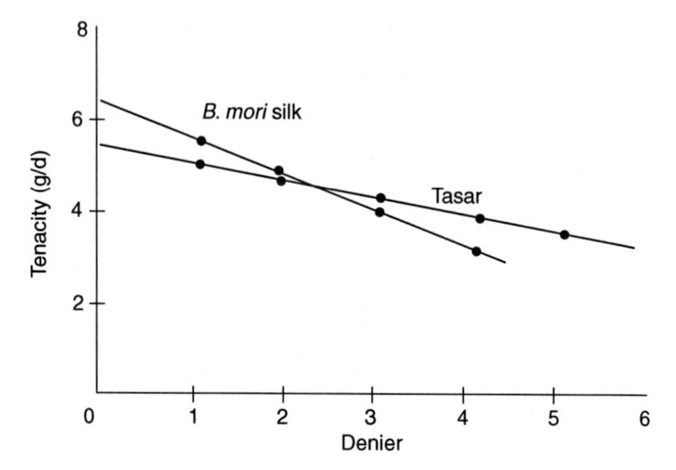

4.1 Tenacity *vs* denier in silk thread.

Table 4.2 Mechanical properties of different varieties of silk

Variety	Sex	Dynamic modulus (10^{10} dyn/cm^2)	Tan δ*	Tenacity (g/d)	Elongation (%)
Shunreix shougetsu	M	1.847	–	5.265	20.36
(mulberry)	F	1.808	–	5.207	21.48
A. mylitta	M	1.132	0.030	3.412	31.36
(tropical tasar)	F	1.087	0.035	3.256	31.12
A. proylei	M	1.305	0.023	4.123	31.45
(temp. tasar)	F	1.087	0.025	4.128	31.48
A. assama (muga)	M	1.205	0.020	3.170	34.83
	F	1.230	0.023	3.823	34.10

* Tan δ is the ratio of the loss modulus to the storage modulus, and is often called the damping or the loss factor.

Table 4.3 LC/SC ratios for different varieties of silk

Group	Variety of silk	Total of short side-chain amino acids (SC)	Total of long side-chain amino acids (LC)	Ratio LC/SC × 100	Tenacity (g/d)	Elongation (%)
B. mori	Mulberry	87.4	11.8	13.5	5.0	24.0
Antheraea silks	Indian tasar	71.1	19.4	27.3	4.1	35.0
	Caligua japonica	68.3	20.5	30.0	1.95	20.0
	Chinese tasar	72.9	16.1	22.1	3.7	35.0

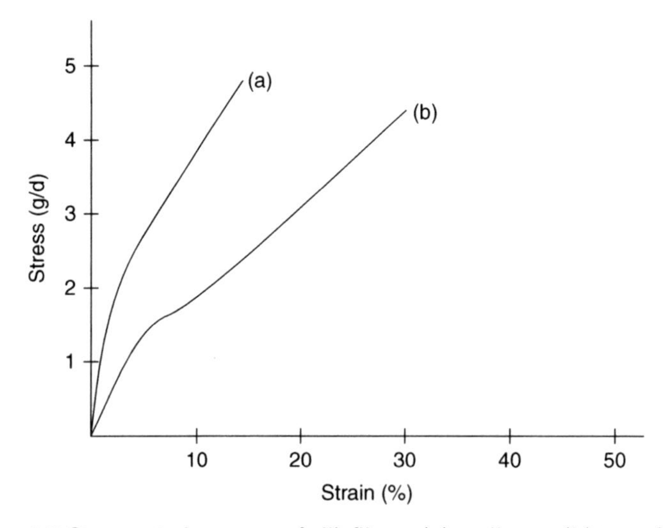

4.2 Stress–strain curves of silk fibres: (a) mulberry; (b) tasar (Sonwalker *et al.*, 1989).

silks. It may be pointed out that the above studies were conducted only on mulberry and tasar. It would be interesting to conduct further studies on two other important silk varieties, muga and eri, with respect to their LC/SC ratio and micro-void content to understand the relation between these parameters and their mechanical properties. Although it is a very good study, it is important to note that while threonine has been considered as short chain, valine, which has a similar residue, has not been considered. These types of analysis need to re-examine for all the varieties of silk.

4.2.1 Stress–strain characteristics

The stress–strain characteristics of a fibre give significant information on its elongation behaviour under the application of stress. The parameters such as initial modulus, yield stress, breaking tenacity, maximum tenacity, breaking load and toughness, can be determined by analysing the stress–strain curves. These parameters can play a significant role in the selection of process parameters for blending and mechanical processing of fibres and designing an end-use specific product.

In a study on tensile behaviour of mulberry and non-mulberry silks (Sonwalker *et al.*, 1989). It was observed that the stress–strain curves followed different patterns for mulberry and non-mulberry silks (Fig. 4.2). Non-mulberry silks such as tasar and muga exhibited a marked yield point in their stress–strain curves which was absent in mulberry silk. It was also observed that a post-yield region existed for tasar and muga and this region was absent in mulberry. In the case of mulberry, the curve exhibited a parabolic nature. In their study on *A.assama* silk, Freddi *et al.* (1994) observed that the stress–strain curve of muga silk single filament demonstrated a high extension at break (40%) and the presence of a marked yield point at an extension of about 5%, followed by a region of gradual extensibility. The behaviour was consistent with that reported by previous workers (Sonwalker *et al.*, 1989). They correlated the relatively high extension of muga silk to unfolding of the long fibroin chains in the amorphous regions.

Meridith (1959) has attributed the differences in the stress–strain behaviour of different varieties of silks to their amino acid composition. Tasar fibres show a relatively high extension after an initial high resistance to extension at low stresses. The slight fold, introduced into the chain molecules of *B. mori* by the presence of small numbers of bulky side-chain amino acids, is probably held in place by weak bonds that help the fibre to resist extension during the initial stressing, so producing a higher initial modulus. The higher proportion of bulky side-chains in tasar fibres result in considerable folding of the molecules within the non-crystalline portions, these folds being held in place by weak cross bonds that rupture at a certain strain level. These linkages will aid resistance to the unfolding of the molecules up to a point, after which rapid extension of the molecular chains can occur by unfolding, until the fibre breaks (Lucas *et al.*, 1955).

The mechanical properties (tenacity, elongation at break and initial modulus) of single filaments were determined by Sen and Murugesh Babu (2004). In their study on some of the Indian silk fibres they compared the tensile properties and stress–strain behaviour of different varities of silk. They also have reported the variations in these properties in different layers of silk filaments within a cocoon. The tenacity, elongation-at-break and initial modulus for different varieties of Indian silk were measured and were found to show interesting differences that provide the basis for the present discussion. It may be observed from Table 4.4 that the average tenacity of mulberry (bivoltine) is 3.75 g/d and that of mulberry (crossbreed) is 3.85 g/d. On the other hand, among the three non-mulberry varieties, tasar shows the highest average tenacity of 4.5 g/d followed by muga (4.35 g/d) and eri (3.7 g/d).

Table 4.4 Average tenacity, elongation and initial modulus values of silk

Variety	Tenacity (g/d)	Elongation-at-break (%)	Initial modulus (g/d)
Mulberry (bivoltine)	3.75	13.55	95.35
Mulberry (crossbreed)	3.85	16.10	106.8
Tasar	4.50	26.50	84.20
Muga	4.35	22.35	81.00
Eri	3.70	20.80	89.05

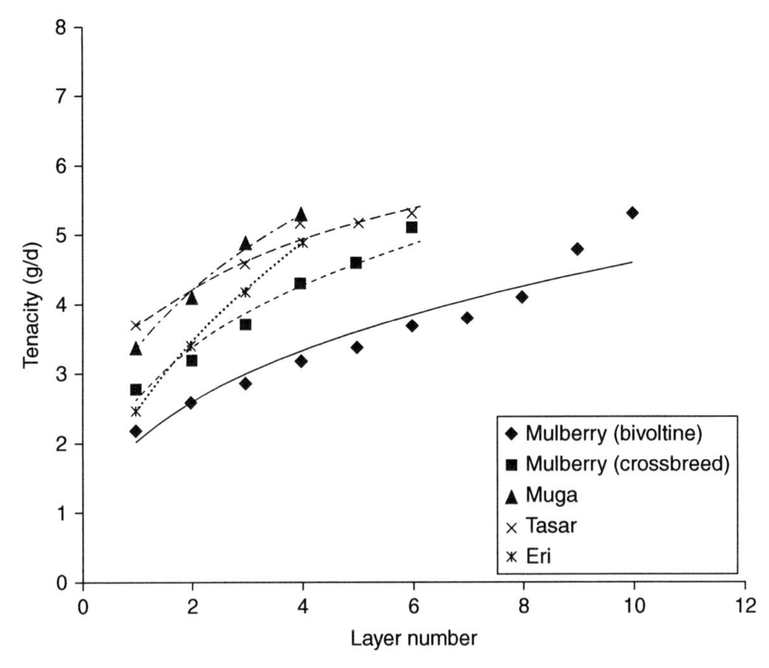

4.3 Tenacity along the filament layers within a cocoon.

It is important to note that the average value of tenacity can be misleading. One may note that the tenacity values increase substantially along the filament length within a cocoon as one moves from the outer to the inner layers. This is true for all the varieties. The tenacity values range from 2.2 to 5.3 g/d (tenacity difference, $\Delta_{ten} \approx 245\%$) for mulberry (bivoltine), 2.8 to 4.9 g/d ($\Delta_{ten} \approx 189\%$) for mulberry (crossbreed), 3.4 to 5.3 g/d ($\Delta_{ten} \approx 155\%$) for muga, 3.7 to 5.3 g/d ($\Delta_{ten} \approx 143\%$) for tasar and 2.5 to 4.9 g/d ($\Delta_{ten} \approx 196\%$) for eri (Fig. 4.3). This means that, depending on the position and size of the test sample, one may obtain values that may be quite different from those expected. The total percentage increase in tenacity may give an impression that mulberry (bivoltine) has the maximum variation along the length, but this could also be a misrepresentation because the filament length in all the varieties is not the same; in fact it is quite different.

Another way to look at this change is to view percentage change per unit length. The percentage increase per 100 m of the filament is around 0.222 for mulberry (bivoltine), 0.286 for mulberry (crossbreed), 0.235 for tasar, 0.325 for muga and 0.445 for eri. This clearly indicates that the variation in mulberry is not all that high, making eri the most non-uniform in terms of tenacity along the fibre. This definitely confirms that, although all the mechanical properties change substantially all along the length, they follow a definite trend. Breaking extension values show a reverse trend compared to that of

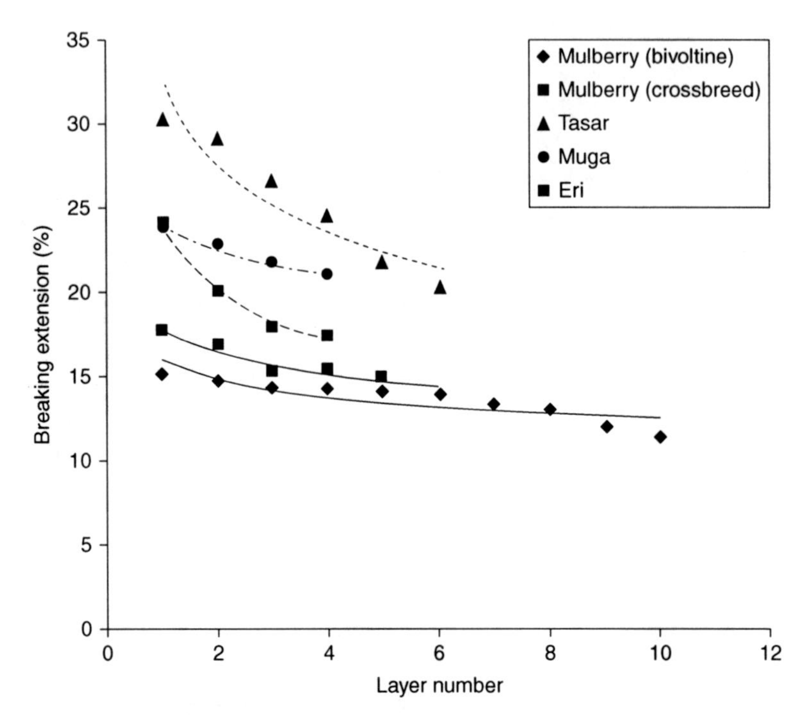

4.4 Breaking extension along the filament layers within a cocoon.

tenacity. Interestingly, higher-breaking extension values were observed for non-mulberry silks. There is a substantial reduction in the extension values along the filament length from the outer to the inner layers. The values for mulberry varieties vary from 15.3% to 11.8% for mulberry (bivoltine) and 17.8% to 14.4% for mulberry (crossbreed). On the other hand, the extension values ranged between 23.8% and 20.9% for muga, 30.4% and 22.6% for tasar and 24.1% and 17.5% for eri (Fig. 4.4).

Two points are worth noting:

1. The reduction in extension values is much less compared to the increase in tenacity.
2. The average extension values of non-mulberry varieties are much higher than those of the mulberry varieties (in all cases high coefficient of variation values ranging from 18.5% to 34.1% were observed).

The relatively high degree of extension in the case of non-mulberry silks may be attributed to the following:

1. All the non-mulberry silks contain more amino acid residues with bulky side groups than the mulberry silk varieties (Iizuka, 1985). This enables

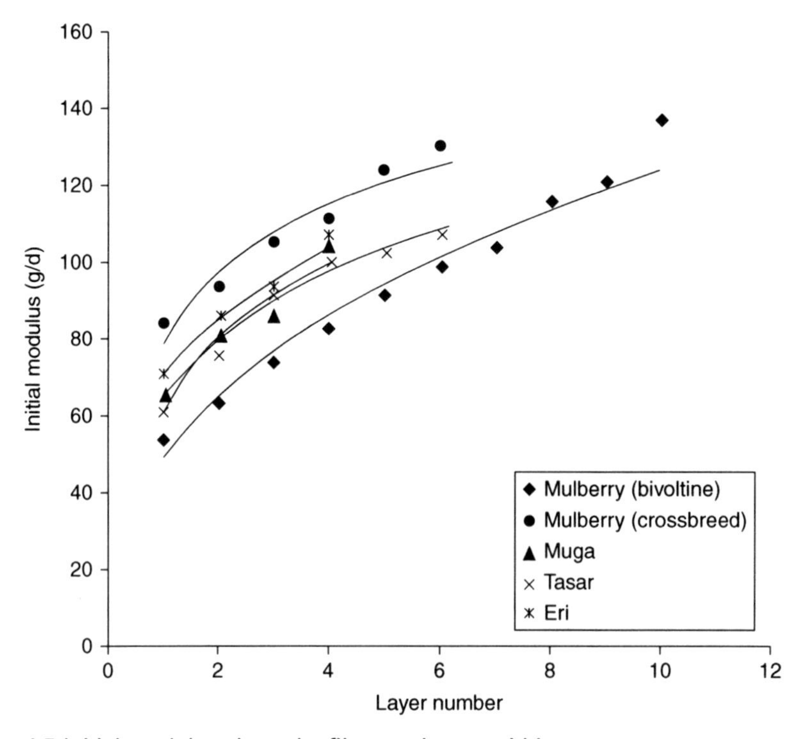

4.5 Initial modulus along the filament layers within a cocoon.

molecular chains in non-crystalline regions in the fibre structure to slip easily when stretched and thus show higher elongation at break.

2. Unfolding of the long fibroin chains in the amorphous regions is a result of either less orientation or less crystallinity (Freddi *et al.*, 1994; Iizuka *et al.*, 1994, 1997).

Another important tensile parameter is the initial modulus. The initial modulus values for different varieties and its change along the filament length are plotted in Fig. 4.5. The initial modulus values follow a similar trend as that of tenacity, showing higher values for the inner layers. The values ranged between 53.9 and 136.8 g/d for mulberry (bivoltine), 83.8 and 129.8 g/d for mulberry (crossbreed), 61.3 and 100.7 g/d for muga, 61.4 and 107 g/d for tasar and 71.1 and 107 g/d for eri silks. This definitely indicates an increase in orientation, in both the crystalline and the amorphous regions, as one moves from the outer to the inner layers.

To gain further insight, it was considered worthwhile to analyse the stress–strain behaviour of these varieties. It may be noted that the curves for both the mulberry silks (Fig. 4.6) do not show any well-defined yield point. In addition, the finer the filament becomes (i.e. as one moves from the outer

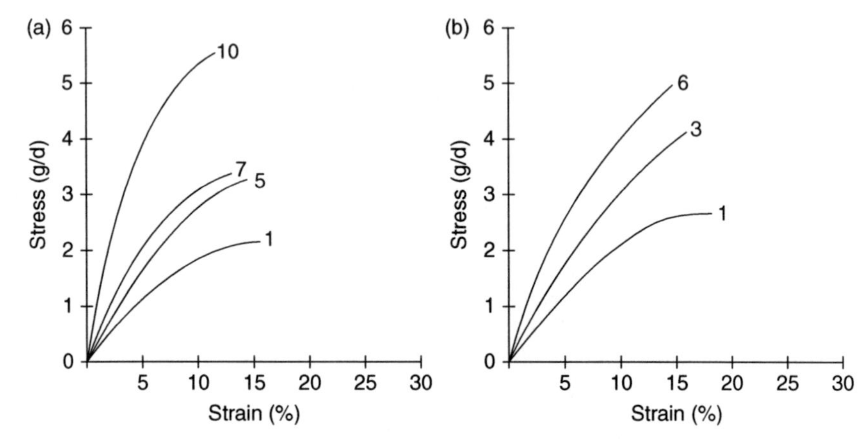

4.6 Stress–strain curves of mulberry silks: (a) bivoltine; (b) crossbreed (NB: The numbers assigned to the curves represent layer numbers).

to the inner layers), the higher the tenacity and modulus and the lower the elongation-at-break. Interestingly, a sharp yield point at a strain level of about 2–3% may be observed for all the non-mulberry silk fibres (Fig. 4.7). Following the yield point, the flow region continues up to 7% strain, depending on the fineness. This region is followed by a strain-hardening region, which was not observed in mulberry silk. This behaviour suggests that mulberry varieties have well-developed crystalline regions that are responsible for smaller and gradual elongation with increasing stress. Tasar, muga and eri, on the other hand, demonstrate a high initial resistance to deformation followed by substantial yielding. This suggests two things:

1. fewer crystalline regions and
2. imperfect crystallites or entanglements, which initially offer resistance but later give way and allow easy deformation in the amorphous regions until strain hardening occurs.

The tensile behaviour of fully degummed filaments of two commercial varieties of silk produced in India, namely mulberry (*B. mori*) and tasar (*A. mylitta*), has been investigated in dry and wet state by Das *et al.* (2005). The mechanical properties of fully degummed mulberry and tasar, in both dry and wet state, are presented in Table 4.5.

It is very interesting to note that for both these filaments, the tenacity and elongation-at-break are not significantly different in dry or wet state. A glance at the typical tensile behaviour reveals that the stress–strain curve of these two varieties is distinctly different, in that tasar shows a clear yield

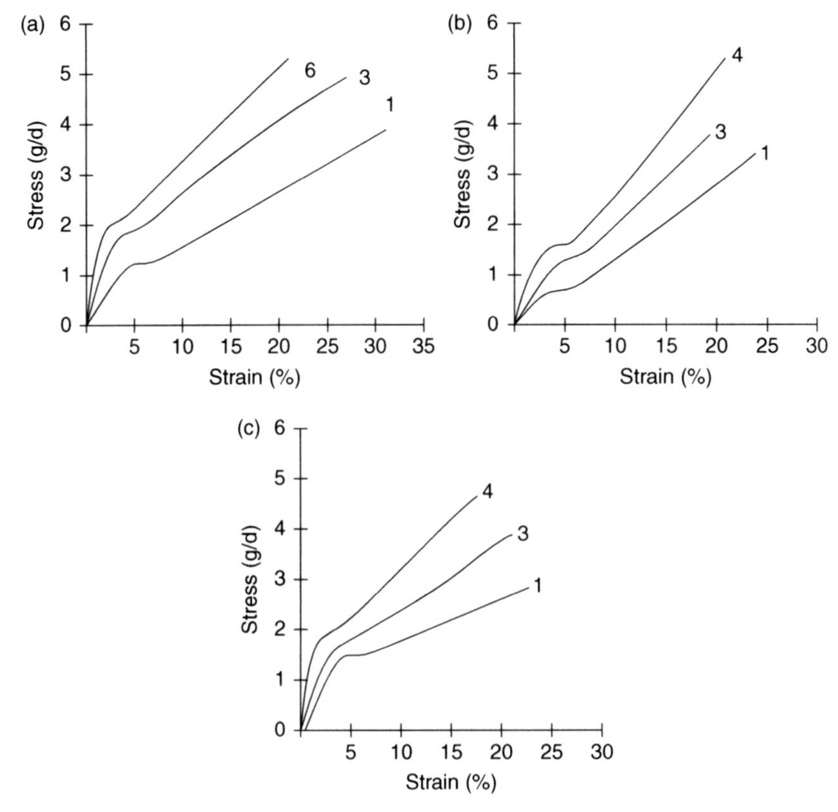

4.7 Stress–strain curves of non-mulberry silks: (a) tasar; (b) muga; (c) eri (NB: The numbers assigned to the curves represent layer numbers).

Table 4.5 Mechanical properties of mulberry and tasar silk in dry and wet conditions

	Mulberry		Tasar	
Properties	Dry	Wet	Dry	Wet
Tenacity (cN/dtex)	1.9 ± 0.2	1.8 ± 0.3	1.6 ± 0.3	1.7 ± 0.3
Elongation-at-break (%)	13.5 ± 3.7	13.3 ± 4.8	29.9 ± 5.9	27.7 ± 6.6
Initial modulus (cN/dtex)	49.5 ± 16.8	42.5 ± 18.6	34.3 ± 5.3	24.5 ± 4.6

point and very high elongation compared to the mulberry filament (Fig. 4.8). This has also been observed by earlier workers. However, wetting does not seem to significantly change the tenacity and extension at break. However, a slight reduction in yield stress is perceptible in both the wetting cases, more so in the case of tasar.

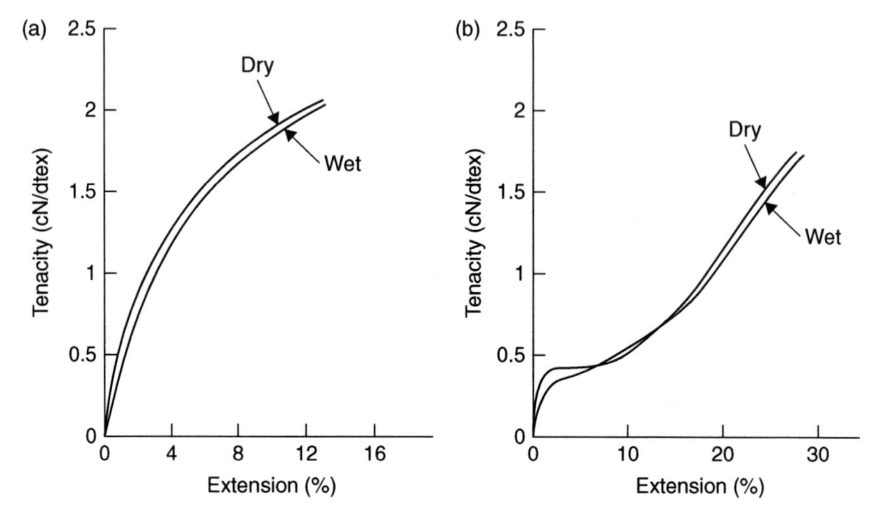

4.8 Stress–strain behaviour of fully degummed silk filaments in dry and wet conditions: (a) mulberry (b) tasar.

4.3 Visco-elastic behaviour

Silk fibre exhibits visco-elastic behaviour. Time-dependent mechanical properties of silk fibres, such as stress relaxation, creep and creep recovery, have also been the subject of interest. Creep is a phenomenon associated with time-dependent extension under an applied load. The complementary effect is stress relaxation under a constant extension. Creep and stress relaxation behaviour of silk has been reported (Das, 1996). The instantaneous extension and secondary creep are both higher for tasar silk compared to those for mulberry silk. The stress relaxation was also found to be more in non-mulberry silks than in mulberry silk.

Silk has also been shown to exhibit the inverse stress relaxation phenomenon (Das, 1996). The inverse relaxation could be observed for both mulberry and tasar silks when the level of strain was maintained below a certain value. Inverse relaxation becomes higher with the increase in peak tension. Cyclic loading has been found to reduce the extent of inverse relaxation.

4.3.1 Creep and stress relaxation

Time-dependent mechanical properties of fibres, such as stress relaxation, creep and creep recovery, have also been the subject of interest. These properties are important, as they would reflect the textile material's behaviour during processing and actual use, particularly the dimensional stability and resiliency. Before a product sees the light of the day, the fibres are subjected

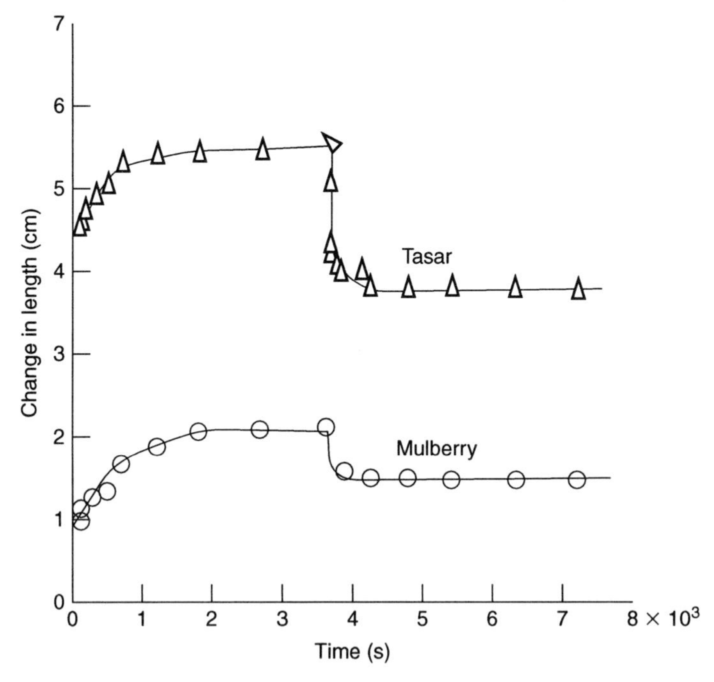

4.9 Creep and recovery of mulberry and tasar.

to a large number of stresses giving rise to internal tensions that relax with time. The consequent variation in dimensions due to above process may cause defects resulting in the rejection of the textile for the application for which it was desired.

Creep is a phenomenon associated with time-dependent extension under an applied load. The complementary effect is stress relaxation under a constant extension. A survey of the literature shows that not much information has been reported on silk. In one study on the creep and stress relaxation behaviour of silk fibres (Das, 1996), Das has reported that instantaneous extension and secondary creep are both higher for tasar silk compared to those for mulberry silk. The stress relaxation was also found to be more in non-mulberry silks than in mulberry silk.

The phenomenon of creep for multivoltine mulberry (*B. mori*) and tropical tasar silk (*A. mylitta*) filaments and relaxation behaviour of mulberry and tasar silk yarn has been reported by Das and Ghosh (2006). The results of creep experiments were plotted and are demonstrated in Fig. 4.9 and Table 4.6. It is observed that the total extension is much larger for tasar than for mulberry. In the total extension, the creep component is more or less similar for mulberry and tasar, that is around 5%. The instantaneous extension is, however, much larger in tasar. As far as the recovery

Table 4.6 Deformation and recovery of mulberry and tasar filament

Variety	Instantaneous extension (%)	Total creep (%)	Total extension (%)	Instantaneous recovery (%)	Primary creep (%)	Secondary creep (%)
Mulberry	5.0	5.5	10.5	2.5	0.5	7.5
Tasar	22.5	5.0	27.5	6.0	2.5	19.0

behaviour is concerned, a great difference is observed especially in the values of secondary creep, which is 7.5% for mulberry and 19% for tasar. The instantaneous recovery and primary creep values are slightly higher in the case of tasar.

Such a difference in the deformation and recovery behaviour can be attributed to the structural differences of the two varieties, that is mulberry and tasar. Tasar has a highly disordered structure manifested by the lower density, birefringence, orientation index and sonic modulus. In addition, tasar has a higher percentage of bulky side groups, as shown by Lucas *et al.* (1995), which will induce easy flow under the application of load. Hence, when such a structure is loaded, both the instantaneous extension and secondary creep are expected to be higher.

4.3.2 Inverse stress relaxation

The term inverse relaxation refers to the building up of tension in a material which has been allowed to recover a part of the extension initially given to it and then constrained to remain at that level of extension (Nachane *et al.*, 1986). It is an important visco-elastic response of the fibre because it reflects the textile material's behaviour during processing and actual use, particularly that related to dimensional stability and resilience. The inverse relaxation or stress recovery, which is a function of time, is initially rapid but decreases gradually with time. The phenomenon of inverse relaxation has been attributed to the visco-elastic nature of a fibre. It may be explained in the following way. A load applied to a visco-elastic material produces immediate elastic and delayed extensions; the former is generally ascribed to the extension of lateral molecular bonds and the latter to molecular slippage resulting from the breakage of some of these bonds. The slippage is time-dependent and goes on until the stress distribution becomes uniform throughout the specimen. Figure 4.10 depicts a typical inverse stress relaxation cycle. When a stretched specimen is allowed to retract, this process is reversed, because the stress now tends to get reduced equally throughout the material and, during the retraction from A to B, the molecular slippage occurs in the reverse direction. Since at point B the specimen is prevented

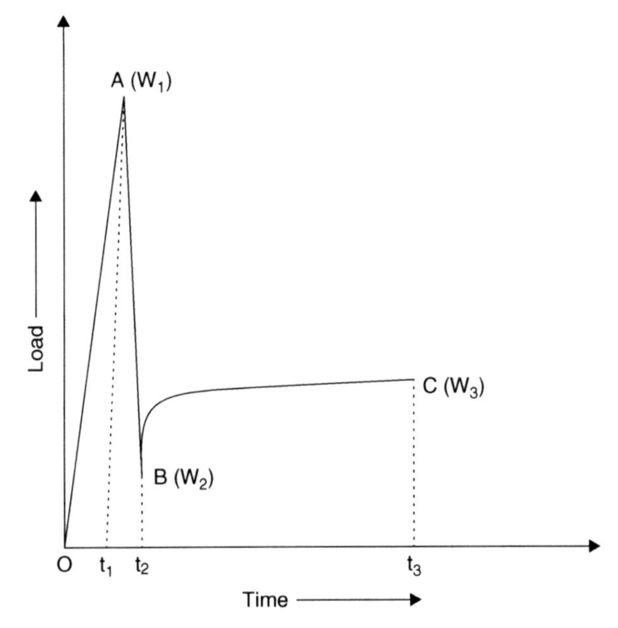

4.10 Typical curve showing inverse stress relaxation behaviour.

from further contraction, the continued molecular process leads to a rise in tension. One may broadly compare this phenomenon to shrinkage stress observed on heating an oriented thermoplastic fibre.

The inverse relaxation has been observed for cotton fibres and yams, ramie, wool, polyethylene terephthalalate (PET) and viscose (Nachane *et al.*, 1982, 1989). Silk has also been shown to exhibit inverse relaxation phenomenon (Das, 1996). In this study, it has been reported that the inverse relaxation could be observed for both mulberry and tasar silks when the level of strain was maintained below a certain value. Inverse relaxation becomes higher with the increase in peak tension. Cycling loading has been found to reduce the extent of inverse relaxation. However, detailed literature on inverse relaxation of silk is not actually available. Typical stress relaxation and inverse stress relaxation behaviours of mulberry (bivoltine) silk (outer layer) are shown in Fig. 4.11. It is interesting to note that, depending on the retraction level of load, the phenomenon represents either the stress relaxation or the inverse stress relaxation phenomenon. It may be observed from these curves that at retraction levels of 0% and 20%, the fibre shows stress relaxation and gives a negative R_i (Inverse Stress Relaxation Index) value (Table 4.7). At retraction levels of 40%, 60% and 80%, however, one observes the inverse stress relaxation phenomenon (i.e. one observes stress build-up). As can be seen from the values given in the tables, the R_i begins with a relatively high negative value at a low retraction level (0% and 20% in the present case, signifying

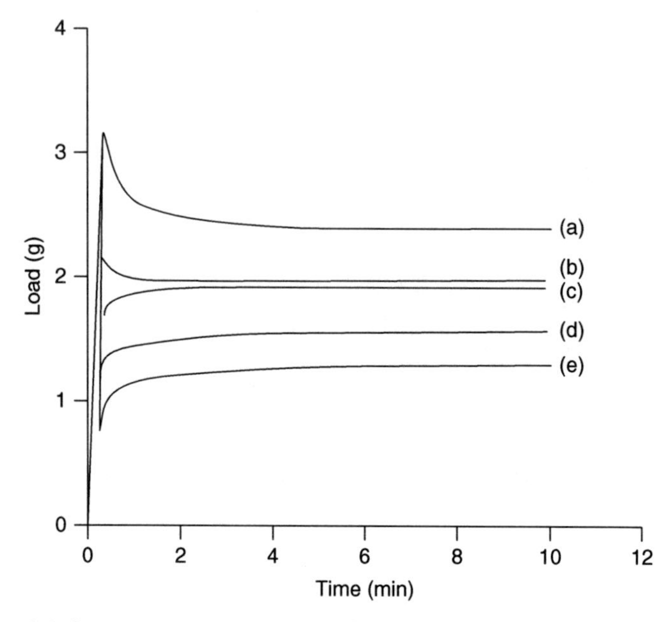

4.11 Inverse stress relaxation of mulberry (bivoltine, outer layer) silk fibres at different levels of retraction: (a) 0%; (b) 20%; (c) 40%; (d) 60%; (e) 80%.

Table 4.7 R_i values for different varieties of silk fibres

	Mulberry (bivoltine)		Mulberry (crossbreed)		Tasar		Muga		Eri	
Retraction level (%)	Outer layer	Inner layer	Outer layer	Inner layer	Outer layer	Inner layer	Outer layer	Inner layer	Outer layer	Inner layer
0	−23.7	−16.1	−25.0	−18.7	−39.0	−33.4	−24.6	−22.6	−28.1	−24.4
20	−7.6	−0.6	−4.0	−3.3	−11.3	−9.4	−11.6	−8.3	−15.0	−10.0
40	7.7	7.0	10.0	8.3	5.4	3.8	6.6	5.8	6.3	5.3
60	8.3	6.7	13.3	10.0	15.0	12.5	11.7	8.0	9.3	7.5
80	16.7	11.7	17.3	12.2	23.8	20.0	21.8	15.0	15.0	11.7

stress relaxation) and becomes increasingly more positive for higher values of retraction, with R_i showing the highest value of 16.7 at a level of 80%. This type of behaviour was observed in all the varieties of silk fibres (Table 4.7). Among the two mulberry varieties, the mulberry (crossbreed) variety shows higher R_i values compared to those of the mulberry (bivoltine) variety. For instance, at an 80% retraction level, mulberry (crossbreed) shows an R_i value of 17.3 (outer layer) as against 16.7 (outer layer) for the mulberry (bivoltine) variety. It is evident from earlier studies that mulberry (crossbreed) shows lower crystallinity and crystallite orientation compared to those of the mulberry (bivoltine) variety. This leads to easy unfolding and molecular relaxation of chains in the amorphous regions, and in general also results in higher stress relaxation as the stress builds up with time.

It is quite clear from Fig. 4.11 that, whether it is stress relaxation or inverse stress relaxation, both involve molecular rearrangement in a way that achieves lower internal energy within the fibre system. The molecular relaxation involves disorientation and coiling up because this state is thermodynamically favourable (i.e. it leads to an increase in entropy), particularly in the amorphous region. The inverse stress relaxation is a process akin to shrinkage. If the fibre is constrained (i.e. not allowed to shrink), then shrinkage stress builds up. As mentioned previously, silk, like any other textile fibre, is a visco-elastic material and undergoes a time-dependent relaxation process. When after stressing to a reasonably high stress level, a sudden decrease in stress level (i.e. retraction) to a level more than it would normally have achieved, in practical time limits, realizes what one may call virtual slackness, and thus may tend to show a shrinkage tendency to achieve thermodynamically stable configuration. In the present case, because the fibre was not allowed to shrink, the stress built up until some equilibrium distribution of stress was achieved.

Figure 4.11 shows a typical curve for mulberry (bivoltine) silk. One would have expected all these curves, those representing stress relaxation or inverse stress relaxation, to theoretically meet at a unique point; that is, at a retraction level where the inversion occurs (Fig. 4.12). However, as one may see this does not happen (Fig. 4.11). The curve at the 0% retraction level ends up at higher stress level compared to that at 20%, and so on. This means that one obtains a band where the final stress level may become constant, depending on the initial stress and the retraction level. Both the mechanical stressing and the retraction lead to some molecular slippage, of course constrained by the crystalline fraction and its nature. This constraint may be called the frictional set that does not allow full relaxation and thus molecular chains can rearrange only to certain permitted levels. This is what is designated as the final stress level. The morphology of the fibre thus has a significant role to play in the relaxation process.

In this respect the non-mulberry silk varieties also behave in a similar manner. First, all these varieties show stress relaxation and inverse stress relaxation phenomena, depending on the retraction level (Table 4.7). The R_i changes from an extreme negative value to a value that is more positive as the retraction level is changed from 0% to 80%. Second, all these varieties show a band where stress levels stabilize, that is the highest stress level for 0% retraction and the lowest for 80% retraction in the present study. It is imperative in all the cases that if retraction levels are further decreased, the stress may stabilize at still lower levels at equilibrium.

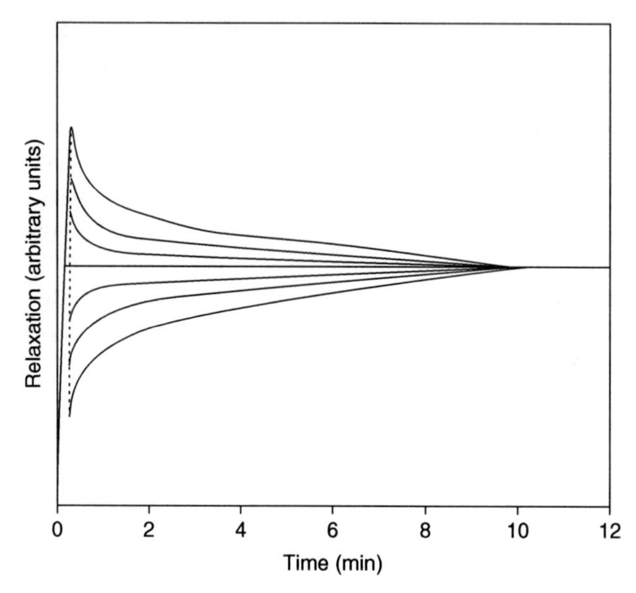

4.12 Typical theoretical curve depicting inverse stress relaxation behaviour.

It is interesting to note, however, that in general the R_i values for non-mulberry varieties are higher than those for mulberry varieties. For instance, at 0% retraction, R_i (%) for tasar is more negative (39.0–33.4) compared to mulberry (23.7–16.1) and at 80% is more positive (i.e. 23.8–20.0), which further substantiates the fact that the lower the morphological order, the higher the likely value of R_i. In the previous study, it was shown that the crystallinity and orientation in non-mulberry varieties are lower (Warwicker, 1956).

It is also interesting to note that R_i values obtained for the outer layers are always greater than those obtained for the inner layers, irrespective of the type of silk. It may be recalled that the microstructural parameters, such as crystallinity, crystallite orientation, birefringence and so forth, show an increase as one moves from the outer to the inner layers, suggesting that the more ordered molecular arrangement in the inner layers does not allow free and easy relaxation of molecules, resulting in lower stress relaxation and stress build-up. Figure 4.13 shows typical curves for mulberry (bivoltine) representing the outer and inner layers, which clearly emphasizes the point mentioned above. This is also clear from the final tension (g) registered for the inner and outer layers (Table 4.8).

Figure 4.12 shows a typical graph that represents the effect of retraction level on the R_i values of tasar fibres. It may be observed that the retraction level at which the R_i becomes zero is identical, at around 32% for the outer

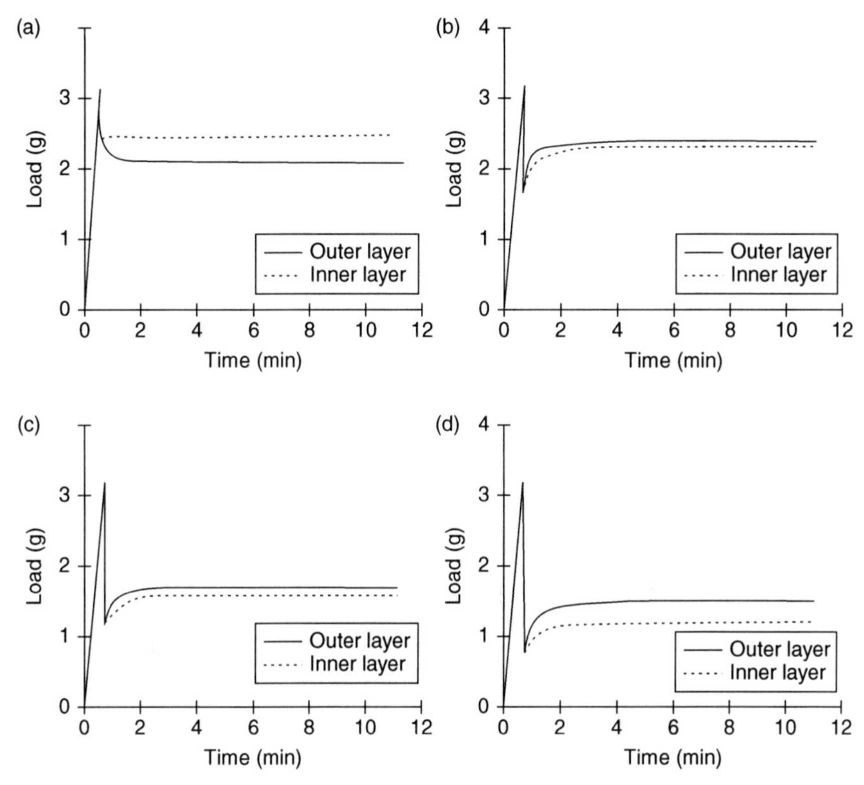

4.13 Inverse stress relaxation of mulberry (bivoltine, outer and inner layers) silk fibres at different levels of retraction. (a) 20% retraction; (b) 40% retraction; (c) 60% retraction; (d) 80% retraction.

Table 4.8 Registered tension (g) after 10 min

	0% Retraction		80% Retraction	
Variety	Outer		Inner	
Mulberry (bivoltine)	2.44	2.70	1.10	0.95
Mulberry (crossbreed)	2.40	2.60	1.10	0.91
Tasar	4.88	5.33	3.20	3.00
Muga	4.52	4.64	2.43	2.00
Eri	2.30	2.42	1.38	1.20

and the inner layers. Similarly, it is around 25% for mulberry (crossbreed), 23–28% for mulberry (bivoltine), 30–32% for muga and 32–33% for eri. Figure 4.14 shows a typical graph that represents the effect of retraction level on the R_i values of tasar fibres. It may be observed that the retraction level at which the R_i becomes zero is identical, at around 32% for the outer

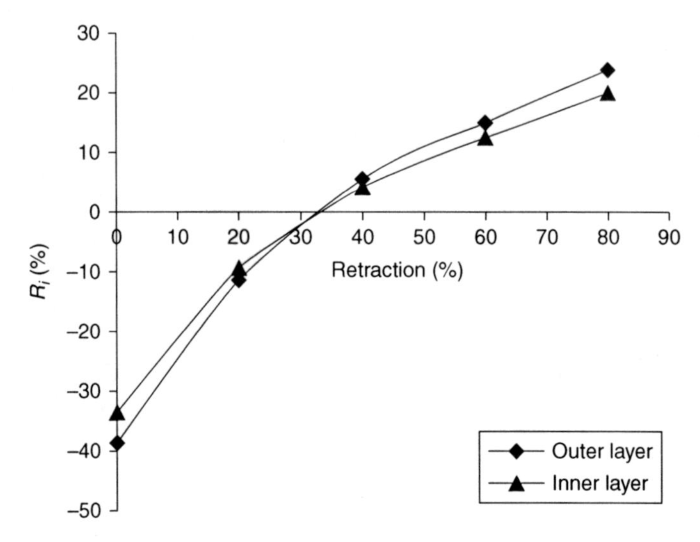

4.14 R_i *vs* retraction level (tasar).

and the inner layers. Similarly, it is around 25% for mulberry (crossbreed), 23–28% for mulberry (bivoltine), 30–32% for muga and 32–33% for eri.

4.4 Dynamic mechanical behaviour

The dynamic mechanical behaviour of different silk fibres in terms of dynamic storage modulus (E′), loss modulus (E″), and the loss tangent tan δ has been investigated by some workers (Magoshi and Magoshi, 1975, 1977; Tsukada *et al.*, 1992, 1994; Freddi *et al.*, 1993, 1995). Figure 4.15 shows the temperature dependence of the dynamic mechanical behaviour of mulberry and tasar fibres. In mulberry, E′ value showed a slight decrease in the temperature range from −10°C to 10°C. It remained almost unchanged as the temperature was increased up to 150°C. The thermal movement of fibroin molecules became evident at above 150°C and the E′ value decreased rapidly at about 170°C. The loss modulus (E″) did not show any significant change until 170°C, at which temperature the broad E″ peak began to appear exhibiting a maximum at 220°C.

It is interesting to note a slight increase of the E′ curve of tasar silk fibres ranging from 30°C to 70°C corresponding to the molecular contraction. Tsukada *et al.* (1992) suggest that there should have been some rearrangements of the tasar fibroin molecules in the amorphous regions during the heating process, resulting in a strengthening of the inter-chain interactions. Mulberry silk did not show this behaviour and their thermal stability below 100°C should be partly attributed to the higher degree of molecular

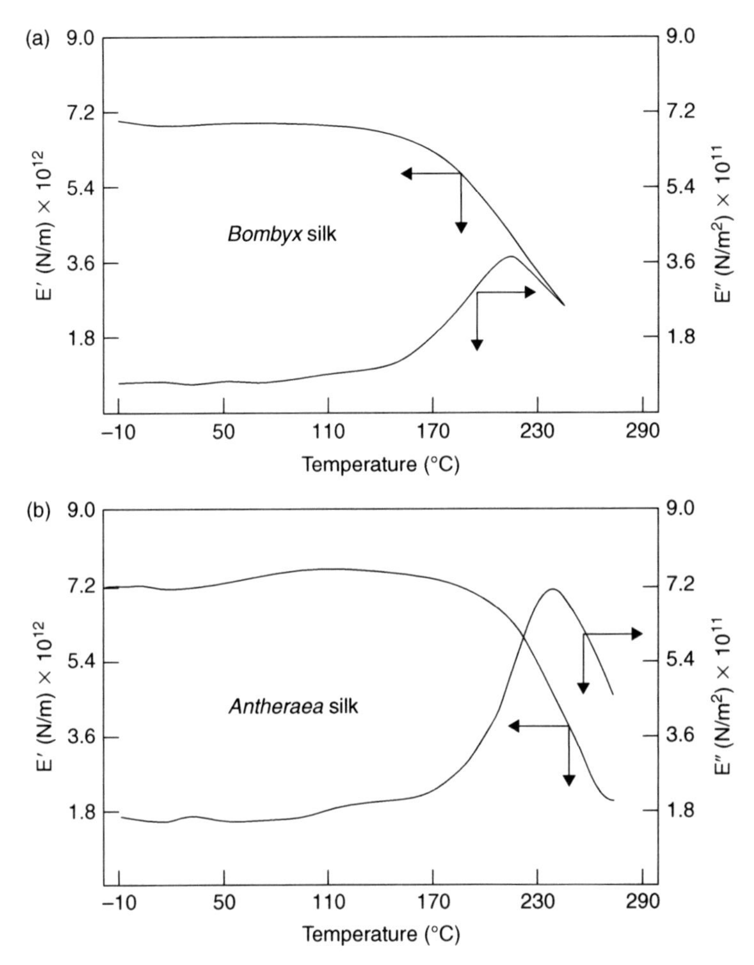

4.15 Temperature dependence of dynamic mechanical behaviour of mulberry (a) and tasar (b) silk fibres (Tsukada, 1992).

orientation even in the amorphous regions as well as to the more compact fibrous structure. The E′ peak appearing at 230° in tasar silk fibres has been attributed to the molecular motion in the crystalline regions because the spacing corresponding to the inter-sheet distance gradually expand at above 190°C at which temperature the E″ peak begins (Freddi *et al.*, 1995).

In a study on dynamic mechanical behaviour for native and regenerated tasar fibres (Fig. 4.16), sharp changes in E′ and E″ values were observed for tasar silk fibres. The changes have been attributed to the segmental motion of fibroin chains in the amorphous regions, becoming more and more intense beyond 160°C, due to inter- and intra-molecular H-bonds breaking and reforming rapidly (Tsukada *et al.*, 1994). Crystalline regions are also

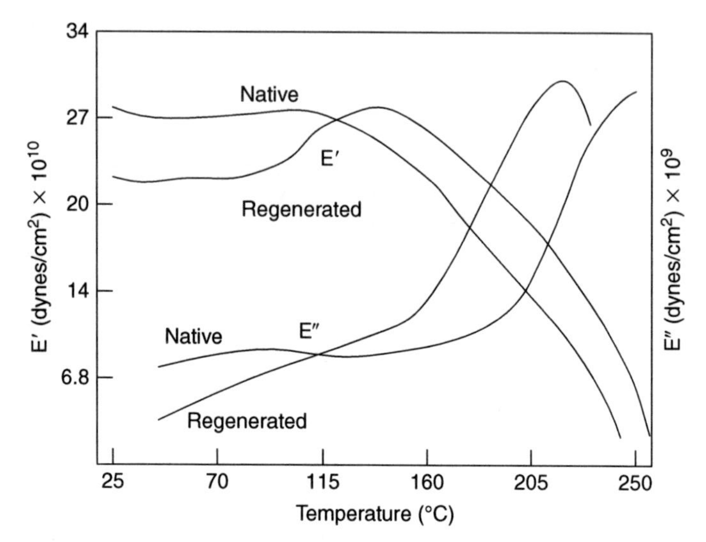

4.16 Dynamic mechanical behaviour of native and regenerated tasar silk fibres (Tsukada *et al.*, 1994).

involved in this process. It has been demonstrated that inter-sheet distance gradually expands above 190°C. The material exhibits a rubber-like behaviour as shown by sharp changes in E′ and E″ values.

4.5 Thermal behaviour

Silk is a non-thermoplastic fibre and is generally not expected to undergo significant morphological changes as a result of thermal shrinkage like synthetic fibres, i.e. polyester, nylon, etc. Silk fibres do undergo several kinds of heat treatments, either in dry or wet state, in the course of textile processing. It is of considerable interest to know if changes due to thermal treatment are significant. In addition, questions such as which of these changes is reversible or not are interesting. The previous discussions indicate that some morphological changes do take place on heating (Tsukada *et al.*, 1992). If true, pre- and post-thermal treatments may affect the textile performances, such as dyeability, mechanical behaviour and handling property.

Studies on the thermal properties of mulberry, eri, muga and tasar conducted using differential scanning calorimetry (DSC), thermogravimetric analysis (TGA), derivative thermogravimetric analysis (DTG), thermomechanical analysis (TMA) and differential thermal analysis (DTA) have been reported by many workers (Tsukada, 1990, 1968, 1969, 1970; Tsukada *et al.*, 1992, 1993a; Talukdar *et al.*, 1991; Bora *et al.*, 1992). Talukdar *et al.* (1991) and Bora *et al.*, (1993) conducted some interesting studies on the Indian silks (mulberry, muga and eri) using DSC, DTA and TGA techniques, over a temperature range of 25–400°C. They observed that the

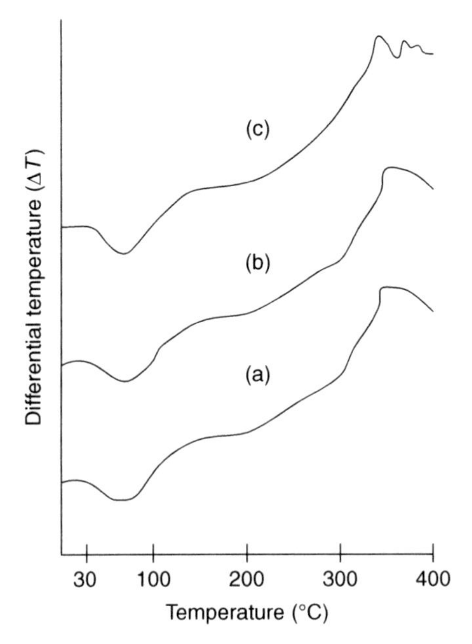

4.17 DTA themograms of silk fibres: (a) mulberry; (b) muga; (c) eri (Talukdar *et al.*, 1991).

thermal transitions in all these fibres occurred almost at the same temperature ranges (Fig. 4.17). They assigned the endothermic peaks at 93°C for muga, 94°C for eri and 93°C for mulberry to the dehydration process. This dehydration stage was evident in the DSC thermograms too (Fig. 4.18). The DSC thermograms of these fibres were marked by the shifting of their baseline representing the change in the heat capacity of the fibres. The DSC traces showed two distinct reactions in two stages. The first endothermic peaks appearing at 150–170°C ($\sim T_{min}$ = 156.2°C) for muga, 70–130°C ($\sim T_{min}$ = 98.3°C) for eri and 79–160°C ($\sim T_{min}$ = 108.5°C) for mulberry respectively may be due to dehydration. The second endothermic peaks occur at the temperature range around 360–370°C ($\sim T_{min}$ = 382°C), 370–410°C ($\sim T_{min}$ = 387°C) and 310–360°C ($\sim T_{min}$ = 304.3°C) for muga, eri and mulberry fibres respectively. These peaks were attributed to the degradation of the fibres caused by the decomposition of the fibrous molecules. It is surprising to note a very sharp endotherm in the case of muga at ~160°C, which the authors have not accounted for (Fig. 4.18). However, according to Tsukada, the sharp endothermic peak may be due to the specific thermal behaviour of calcium oxalate present in the sample (Tsukada, 1982).

The TGA curves (Fig. 4.19) showed the beginning of weight loss at 58°C for muga, 46°C for eri and 49°C for mulberry. The process became rapid at about 220°C for all the fibres. The thermal decomposition occurred beyond 260°C, 255°C and 240°C for muga, eri and mulberry respectively. At about

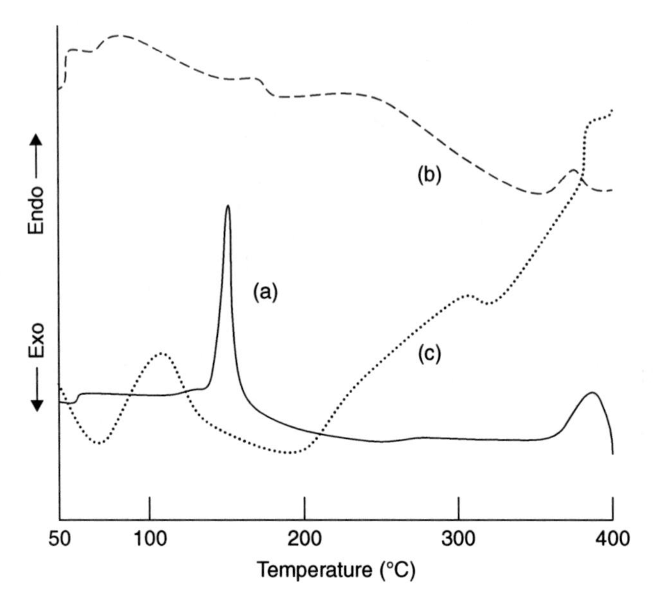

4.18 DTA themograms of silk fibres in air atmosphere: (a) muga; (b) eri; (c) mulberry (Talukdar *et al.*, 1991).

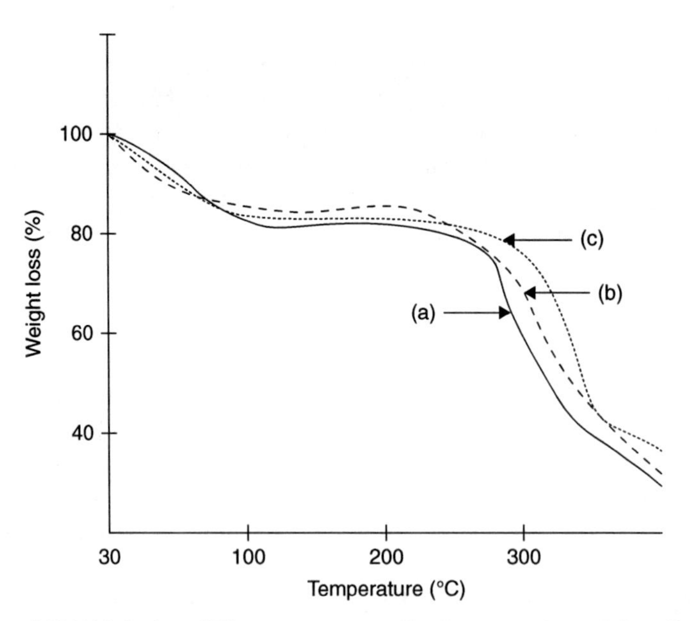

4.19 Weight loss (%) *vs* temperature in air atmosphere: (a) mulberry; (b) muga; (c) eri (Bora *et al.*, 1992).

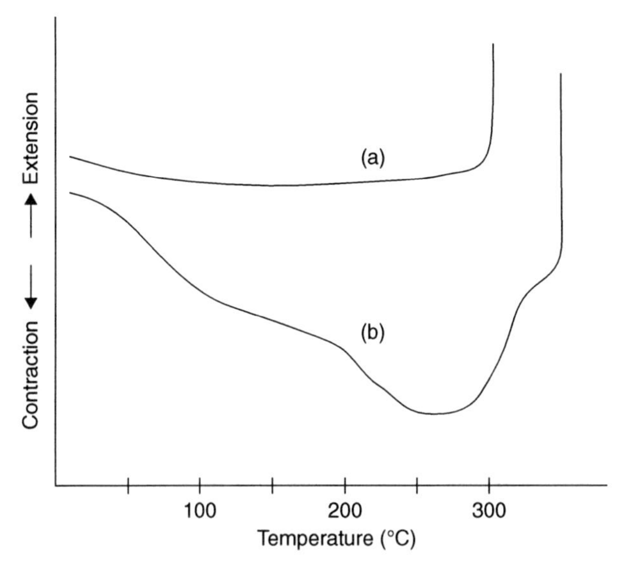

4.20 TMA thermograms of silk fibres: (a) *B. mori* (mulberry); (b) *A. pernyi* (tasar) (Tsukada *et al.*, 1992).

400°C muga, eri and mulberry silks showed a weight loss of about 37%, 49% and 42% respectively. From their kinetic data and the thermodynamical properties, Bora *et al.* (1992) concluded that the first transition points of all the thermo grams of the hygroscopic silk fibres represent the dehydration processes. These are governed by a mechanism involving dissociation of the water molecules mostly embedded in the amorphous region of the semicrystalline fibres. The second transition points of all the thermo grams represent thermal decomposition and degradation of crystalline set-up of the silk fibres. These studies tend to show that the thermal behaviour of all the silks are more or less the same (Fig. 4.19), which is difficult to understand as the chemistry of mulberry and non-mulberry are quite different.

In an interesting study on the thermo-mechanical behaviour of silks, Tsukada *et al.* (1992) have reported the thermal expansion and contraction properties of mulberry and tasar (*A. pernyi*) silk fibres measured by thermomechanica1 analysis (TMA). In this study, they observed that (Fig. 4.20) in the range from room temperature to 120°C, the *B. mori* fibre exhibited a slight contraction of about 0.7% which was attributed to the evaporation of water absorbed by the specimen. Predominant extension of the fibre was observed at around 310°C, due to the breaking and reforming of the inter-chain hydrogen bonds and to the partial thermal decomposition. On the other hand, tasar silk fibres exhibited a prominent two-step contraction in the temperature range from 25°C to 250°C. The first step appeared in the temperature range from 25°C to about 100°C, and the second step started at above 200°C. This was accompanied by the abrupt change in slope of the TMA curve, and attained a maximum at 250–260°C. The first rapid extension occurs at above 260°C and the final abrupt extension occurred at about 340°C, 30°C higher than the *B. mori* silk fibres. These studies concluded that the tasar silk decomposes at higher temperature due to its thermal stability of the $-(Ala)_n-$sequences forming the crystalline regions. However, the maximum amount of contraction was exhibited by tasar in the heating process, which was about 3.6%, more than five times higher than that of *B. mori* fibres.

Other thermal studies conducted on mulberry silk (Hirabayashi *et al.*, 1975; Magoshi and Magoshi, 1977; Magoshi *et al.*, 1977; Tsukada *et al.*, 1978; Agarwal *et al.*, 1997) and other wild silks clearly indicate that the broad endotherm below 100°C is due to the dehydration process and that non-mulberry silks are stable up to 200°C. Bora *et al.* (1993) showed that tasar exhibited two minor and broad endothermic peaks (shoulder form), one at about 234°C and another at around 296°C. On the other hand, mulberry silks showed a single endothermic peak at about 320°C. The tasar silk fibres showed a major endothermic peak at 362°C which was attributed to the thermal decomposition of the fibres. The minor endotherms occurring above T_g (190–200°C) were suggested to be related to the molecular motion of the fibroin chains either in the amorphous and laterally ordered regions or in the crystalline regions. Two minor and broad endothermic transitions at 230°C and 300°C have been reported which are followed by a prominent endothermic peak at 362°C that has been attributed to the thermal decomposition of silk fibres.

Thermal analysis is a useful tool for monitoring important processing parameters and end-use properties of textile fibres. In addition to other factors, the response of a fibrous polymer to a thermal treatment may depend on its chemical architecture and microstructure. The thermal transitions may have a significant effect on the processing conditions. These may also give directions for affecting modifications in the fibres to help improve their performance, such as dyeability, mechanical behaviour and handle.

It is of considerable interest to know if thermal treatments bring about any morphological changes in silk fibres that will have a bearing on the mechanical and other properties. In addition, questions like which of these changes is reversible or not are very interesting. Some of the earlier studies indicated that heat treatments do bring about changes in the amorphous and crystalline regions. Magoshi and Nakamura (1975) and Magoshi and Magoshi (1977) studied the DSC thermograms of amorphous silk fibroin of *B. mori*. They observed one endothermic shift at 175°C, which they assigned to the glass transition, one exothermic peak at 212°C due to crystallization and a peak at 280°C related to degradation. On the other hand, the amorphous tussah silk fibroin showed an endothermic shift at 162°C, an endothermic peak at 220°C due to α–β transition and an exothermic peak at 230°C due to crystallization

(Magoshi and Magoshi, 1977; Magoshi *et al.*, 1977). The DSC curves of tussah silk fibres (Tsukada, 1986; Tsukada *et al.*, 1992; Agarwal *et al.*, 1997) exhibited two minor and broad endothermic shoulders at about 234°C and at 296°C and a major endothermic peak at 362°C that was attributed to the thermal decomposition of these fibres. *B. mori* silk fibre, however, showed a single endothermic peak at about 320°C.

Tsukada *et al.* (1978) investigated the thermal behaviour of *B. mori* silk fibre by DTA under different nitrogen pressures. They reported a single endothermic peak at 319°C under the pressure of 1.0 kg/cm^2. When the nitrogen pressure was increased to 61 kg/cm^2, two endothermic peaks were observed, one at 300°C and the other at 326°C. Talukdar *et al.* (1991) and Bora *et al.* (1993) have studied the three Indian varieties of silk fibres, mulberry, muga and eri by DTA and TGA. The DTA curves exhibited endothermic peaks at about 93°C for all the three silk fibres, which were followed by an exothermic peak at around 312°C for mulberry and muga and about 315°C for eri. From the TGA analysis, it was found that mulberry exhibited maximum weight loss of about 52%, followed by eri (49%) and muga (37%).

Similarly, Freddi *et al.* (1994) investigated the thermal properties of Indian muga silk fibres using DSC. According to this study, muga exhibited a broad endotherm below 100°C due to the evaporation of water. Two endothermic shoulder forms were observed at around 230°C and 300°C, followed by a prominent endothermic peak at 362°C. There are indications that the mulberry and non-mulberry silk undergoes thermal transitions in a different manner. Although many studies have been conducted on mulberry and tasar silk fibres, the Indian non-mulberry silk fibres such as muga and eri need to be studied more carefully as the availability of these silks is scarce and these varieties are very sensitive to thermal inputs.

In an interesting study the thermal behaviour of different varieties of silks have been reported (Murugesh Babu and Sen, 2007). Mulberry and non-mulberry silks were subjected to TGA and DTA studies. The TGA curves of mulberry (bivoltine) silk variety recorded in air and in nitrogen atmospheres is shown in (Mulberry, crossbreed variety shows the same trend) and the results are summarized in Table 4.9. It may be observed from the TGA curves that both varieties are stable up to about 200°C. Only around 4% of the weight loss is observed until this temperature. The weight loss starts increasing at around 220°C. The weight loss at 400°C observed for mulberry (bivoltine) in air atmosphere is lower (about 42%) as compared to about 44% for mulberry (crossbreed) varieties (Table 4.9). However, it may be seen from the TGA thermograms that the weight loss at 400°C in nitrogen is marginally higher at around 46% for mulberry (bivoltine) and around 45% for mulberry (crossbreed) (Table 4.9). Does this mean more degradation is occurring in nitrogen atmosphere? It seems unlikely. The degradation is generally accompanied by oxidation reaction. The addition of oxygen can

Table 4.9 Results of TGA analysis (air atmosphere)

Variety	Temperature of initiation (°C)	Temperature of noticeable change in weight (°C)	Temperature of onset of rapid degradation (°C)	Temperature of inflexion (°C)
Mulberry (bivoltine)	220	262	318	342
Mulberry (crossbreed)	218	263	315	348
Tasar	220	285	365	395
Muga	220	285	357	395
Eri	220	285	358	395

to some extent offset the weight loss in air atmosphere resulting in higher weight retention. It may be noted that the weight loss is less than 50% even at 400°C and less than 4% at 100°C and 200°C. This goes on to substantiate that as such mulberry silk is stable to heat and at normal processing temperatures of dyeing (≤100°C) or ironing (≈150–170°C), silk fabric should not face any difficulties on account of thermal inputs, as long as these are of practically reasonable duration. From the data of the weight loss in air (Table 4.10), it is clear that in the case of mulberry silk, the initiation of degradation process starts somewhere around 220°C which becomes noticeable at around 262°C, and goes through a rapid change at around 315°C and the rate shows a distinct reduction after 340°C. This is also evident from the differential curves (Fig. 4.21).

TGA curves of eri (muga and tasar behave similarly) are shown in Fig. 4.22. Here too, the initial part of the curves is stable (up to about 200°C), demonstrating a weight loss of only about 5% (Table 4.10). As the temperature increases, the weight loss begins to increase from 220°C onwards. It is interesting to note that at 400°C, the weight loss percentage observed for these varieties in air atmosphere is much lower than at about 37% for tasar, 36% for muga and 35% for eri as compared to mulberry varieties (41–44%). This may be attributed to: (i) the presence of more electrostatic linkages due to higher amount of acidic and basic amino acid residues or (ii) the formation of inter- or intra-molecular anhydride or amide linkages due to availability of more susceptible groups. This may cause delay in the degradation process or increase in degradation temperature. Earlier studies have shown that the ratio of bulky/non-bulky amino acids in non-mulberry silk is higher (0.24–0.32) as compared to mulberry varieties (0.17–0.18) (Sen and Babu, 2004), and also (iii) to higher thermal stability of non-mulberry silk due to the presence of the $-(Ala)_n-$ sequences in the crystalline regions (Tsukada *et al.,* 1987, 1992). As is the case with mulberry, a significant

Table 4.10 Weight loss (%) of different varieties of silk at different temperatures

	Air atmosphere			Nitrogen atmosphere		
Variety	100°C	200°C	400°C	100°C	200°C	400°C
Mulberry (bivoltine)	2.0	4.0	41.9	2.0	4.0	46.4
Mulberry (crossbreed)	2.0	4.0	44.4	2.0	4.0	45.0
Tasar	3.0	5.0	37.7	3.0	5.0	42.2
Muga	2.0	4.0	36.7	2.0	4.0	40.6
Eri	2.0	4.0	35.4	2.0	4.0	45.1

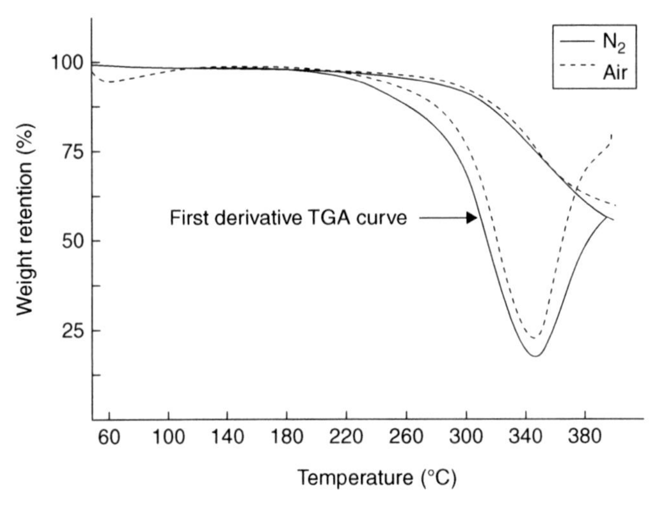

4.21 TGA thermograms of mulberry (bivoltine) silk fibres.

increase in weight loss is also observed in nitrogen atmosphere for these varieties. The increase is about 4–5% for tasar and muga whereas it is about 10% for eri. This increase in the weight loss in nitrogen atmosphere is, however, higher as compared to mulberry varieties (1–3%). The degradation profile as seen from the weight loss (%) curves is quite different up to 400°C. While both mulberry varieties show a sigmoidal curve with inflexion point at about 342°C (Fig. 4.22), the non-mulberry varieties begin to show the sign of inflexion at about 395°C. Similarly, although the initiation of weight loss is seen around the same temperature of 200°C, the noticeable fall in weight starts at around 280°C (≈20°C higher than mulberry), which becomes rapid at around 360°C (≈40°C higher than mulberry). All these observations point towards the higher thermal stability of the non-mulberry silk. It has been reported earlier (Sen and Babu, 2004) that the non-mulberry varieties

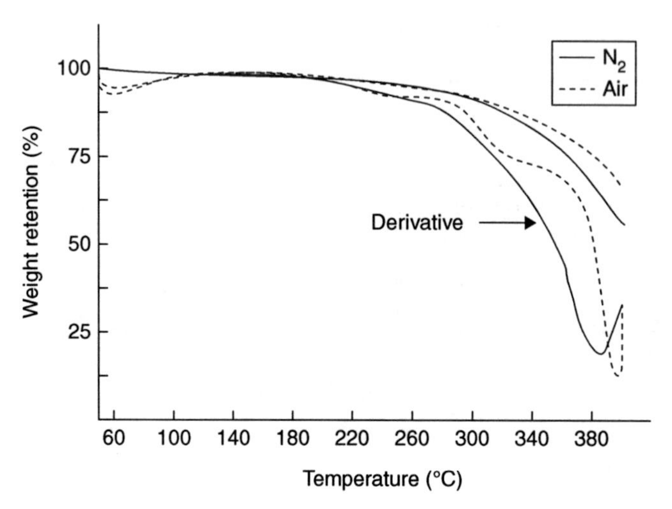

4.22 TGA thermograms of eri silk fibres.

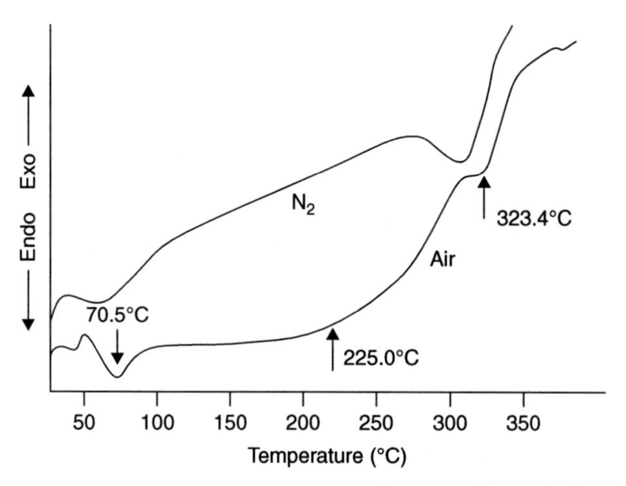

4.23 DTA thermograms of mulberry (bivoltine) silk fibres.

of silk have higher molecular weight compared to the mulberry varieties. This may also contribute, to some extent, to the higher thermal stability of non-mulberry varieties.

The DTA curves of mulberry silk fibres measured from room temperature to 400°C in air and in nitrogen atmospheres are represented in Fig. 4.23. Various transitions are tabulated in Table 4.11. Both thermograms exhibit a first minor endotherm below 100°C (≈62°C). This may be attributed to the evaporation of water. It may be noticed that these fibres show noticeable thermal stability up to 280°C, as evident from thermograms obtained in nitrogen

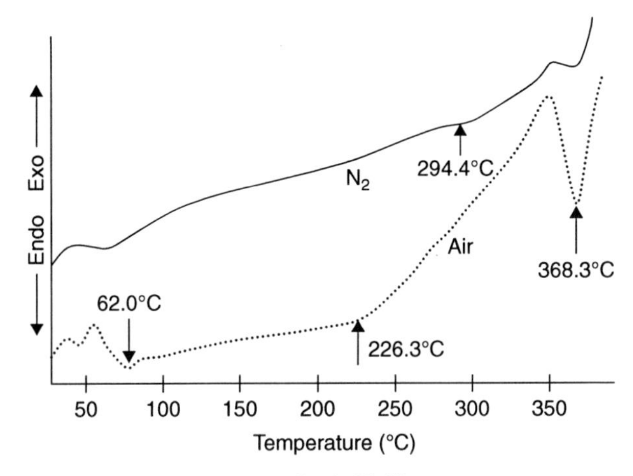

4.24 DTA thermograms of eri silk fibres.

Table 4.11 Transition temperatures (°C) observed in DTA thermograms

	Transition 1		Transition 2		Transition 3		Transition 4	
Variety	Air	N_2	Air	N_2	Air	N_2	Air	N_2
Mulberry (bivoltine)	70.5	63.6	225.0	–	–	–	323.4	318.9
Mulberry (crossbreed)	69.6	67.2	226.0	–	–	–	322.6	315.3
Tasar	69.5	63.6	225.2	229.6	–	294.6	360.5	362.4
Muga	68.4	66.4	225.7	230.9	–	296.1	365.0	369.4
Eri	62.0	61.8	226.3	232.1	–	294.4	368.3	370.3

atmosphere. A major prominent endothermic peak appears at around 320°C. This is attributed to the thermal decomposition of fibres. The TGA thermograms also show a distinct increase in the weight loss around this temperature. The DTA thermograms of the mulberry silk in air show almost no reactions up to ≈220°C whereafter definite exothermic changes are suggested; it becomes rapid until about 310°C where a plateau suggests a balance of exothermic (oxidation) and endothermic reactions (molecular chain breaking) until about 320°C, whereafter rapid exothermic reactions are seen that continue until the end of the experiment (400°C). However, interestingly when this experiment was performed in nitrogen, one does not observe any sharp change in slope up to 280°C, whereafter an endothermic peak is observed at around 315°C signifying molecular degradation, followed by a rise in the exothermic reactions. It is but natural that oxidative processes are restrained until a temperature after which major breakdown takes place.

DTA thermograms of eri fibres are represented in Fig. 4.24 (muga and tasar behave similarly). The non-mulberry silk also shows three major thermal transactions, one around 70°C due to the evaporation of water, the second around 225°C representing the start of oxidative reactions, and then at 365°C signifying major degradative reactions. The actual temperatures for different varieties are listed in Table 4.11. These results are in agreement with the earlier TGA results. In nitrogen, the oxidative processes are restrained until major breakdown takes place at around 360°C as expected. Thereafter exothermic reactions are rapid. However, an additional transition at around 290°C is clearly observed with the non-mulberry silk suggesting molecular rearrangement, thereby compensating the exothermic process. This, however, needs to be further investigated.

4.6 References

Agarwal, N., Hoagland, D.A. and Farris, R.J. (1996), Effect of moisture absorption on the thermal properties of *Bombyx mori* silk fibroin films, *J. Appl. Polym. Sci.*, **63**, 401–410.

Bora, M.N., Baruah, G.C. and Talukdar, C.L. (1992), Investigation on the thermodynamical properties of some natural silk fibres with various physical methods, *Thermodynamica Acta*, **218**, 425–434.

Das, S. (1996), Studies on tasar silk, PhD thesis, IIT, Delhi.

Das, S., Chattopadhyay, R., Gulrajani, M.L. and Sen, K. (2005), Study of property & structural variants of mulberry and tasar silk filaments, *AUTEX Res. J.*, **5**(2),

Das, S. and Ghosh, A. (2006), Study of creep, stress relaxation, and inverse relaxation in mulberry (*Bombyx mori*) and tasar (*Antheraea mylitta*) silk, *J. Appl. Polym. Sci.*, **99**, 3077–3084.

Freddi, G., Gotoh, Y., Mori, T., Tsutsui I. and Tsukada M. (1994), Dyeability of silk fabrics modified with dibasic acid anhydrides, *J. Appl. Polym. Sci.*, **52**, 775–781.

Freddi, G., Romano, M., Rosaria, M. and Tsukada, M. (1995), Silk fibrion/cellulose blend films: Preparation, structure, and physical properties, *J. Appl. Polym. Sci.*, **56**, 1537–1545.

Freddi, G., Svilokos, A.B., Ishikawa, H. and Tsukada, M. (1993), Chemical compositionand physical properties of Gonometa rufoburnae silk, *J. Appl. Poly. Sci.*, **48**, 99–106.

Hardy, J.G., Romer, L. and Scheibel, T. (2008), Polymeric materials based on silk proteins, *Polymer*, **49**, 4309–4327.

Hirabayashi, K., Tsuakada, M., Nakura, M. and Ishikawa, H. (1975), The change of thermal characteristics of tussah silk fibroin with drawing, *Sen-I Gakkaishi*, **31**, 1.

Iizuka, E. (1985), Silk thread-mechanism of the spinning and the physical properties, *J. Appl. Polym. Sci. Appl. Polym. Symp.*, **41**, 173.

Iizuka, E. and Itoh, H. (1997), Physical properties of eri silk, *Int. J. Wild Silkmoth Silk*, **3**, 37.

Iizuka, E., Kawano, R., Kitani, Y., Okachi, Y., Shimizu, M. and Fukuda, A. (1993a), Studies on the physical properties of Indian non-mulberry silks: I. *Antheraea proylei*, *Indian J. Seric.*, **32**, 27.

Iizuka, E., Okachi, Y., Shimizer, M., Fukuda, A. and Hashizume, M. (1993b), *J. Seric.*, **1**, 1.

Iizuka, E., Okachi, Y., Shimizer, M., Fukuda, A. and Hashizume, M. (1993c), *Indian J. Seric.*, **32**, 175.

Iizuka, E., Vegaki, K., Takamatsu, H., Okachi, Y. and Kawai, E. (1994), *J. Seric. Sci. Jpn.*, **63**, 64.

Lucas, F., Shaw, J.T.B. and Smith, S.G. (1955), The chemical constitution of some silk fibroins and its bearing on their physical properties, *J. Text. Inst.*, **46**, T440–T452.

Magoshi, J. and Magoshi, Y. (1975), Physical properties and structure of silk. II. Dynamic mechanical and dielectric properties of silk fibroin, *J. Polym. Sci.*, **13**, 1347–1351.

Magoshi, J. and Magoshi, Y. (1977a), Physical properties and structure of silk. V. Thermal behavior of silk fibroin in the random-coil conformation, *J. Polym. Sci.*, **15**, 1675–1683.

Magoshi, J. and Magoshi, Y. (1977b), Physical properties and structure of silk. III. The glass transition and conformational changes of tussah silk fibroin, *J. Polym. Sci.*, **21**, 2405–2407.

Magoshi, J., Magoshi, Y., Nakamura, S., Kasai, N. and Kakodo, M. (1977), *J. Appl. Polym Sci.*, **15**, 1675–1683.

Meridith, R. (1959), *The Mechanical Properties of Textile Fibres*. North Holland Publishing Company, Amsterdam.

Murugesh Babu, K. and Sen, K. (2007), Thermal behavior of silk, *Res. J. Text. Apparel*, **11**(2), 21–27.

Nachane, R.P., Hussain, G.F.S. and Krishna Iyer, K.R. (1982), Inverse relaxation/stress recovery in cotton fibers and yarns, *J. Appl. Polym. Sci.*, **52**, 483–484.

Nachane, R.P., Hussain, G.F.S., Patel, G.S. and Krishna Iyer, K.R. (1986), Inverse relaxation in spun yarns, *J. Appl. Polym. Sci.*, **31**, 1101–1110.

Nachane, R.P., Hussain, G.F.S., Patel, G.S. and Krishna Iyer, K.R. (1989), A study of inverse relaxation in some textile fibers, *J. Appl. Polym. Sci.*, **38**, 21–27.

Scheibel, T. (2004), Spider silks: recombinant synthesis, assembly, spinning, and engineering of synthetic proteins, *Microb. Cell Fact.*, **3**, 14.

Scheibel, T. (2005), Protein fibers as performance proteins: new technologies and applications, *Curr. Opin. Biotechnol.*, **16**, 427–433.

Sen, K. and Babu, K.M. (2004), Studies on Indian Silk – I: Macrocharacterization and analysis of amino acid composition, *J. Appl. Polym. Sci.*, **92**(2), 1080–1097.

Sonwalker, T.N., Roy, S., Vasumathi, B.V. and Hariraj, G. (1989), *Ind. J. Seric.*, **2**, 159–163.

Talukdar, C., Baruah, G.C. and Bora, M.N. (1991), Infrared spectroscopic study of some natural silk fibres, *Ind. J. Phys.*, **65B**(6), 641–649.

Tsukada, M. (1982), *J. Seric. Sci. Jpn.*, **51**, 499–502.

Tsukada, M. (1990), Characterization of methacrylonitrile-grafted silk fibers, *J. Appl. Polym. Sci.*, **39**, 1289–1297.

Tsukada, M., Freddi, G., Minoura, N. and Allara, G. (1994), Preparation and application of porous silk fibroin materials, *J. Appl. Polym. Sci.*, **54**, 507–514.

Tsukada, M., Freddi, G., Nagura, M., Ishikawa, H. and Kasai, N. (1992), Structural changes of silk fibers induced by heat treatment, *J. Appl. Polym. Sci.*, **46**, 1945–1953.

Tsukada, M., Freddi, G., Shiozaki, H. and Pusch, N. (1993a), Changes in physical properties of methacrylonitrile-grafted silk fibers, *J. Appl. Polym. Sci.*, **49**, 593–598.
Tsukada, M., Freddi, G., Monti, P. and Bertoluzza, A. (1993b), Physical properties of silk fibers grafted with a binary mixture of styrene and n-butyl methacrylate, *J. Appl. Polym. Sci.*, **49**, 1565–1571.
Tsukada, M., Goto, Y., Freddi, G. and Shiozaki, H. (1992), Chemical modification of silk with aromatic acid anhydrides, *J. Appl. Polym. Sci.*, **45**, 1189–1194.
Tsukada, M., Hirabayashi, K., Komoto, T. and Kawai, T. (1978), *Sen-I Gakkaishi*, **34**, 8.
Warwicker, J.O. (1956), The crystal structure of silk fibroins, *Trans. Faraday Soc.*, **52**, 554–557.

5
The dyeing of silk

DOI: 10.1533/9781782421580.117

Abstract: This chapter discusses the dyeing of silk, including types of dye used for silk, factors affecting dyeing behaviour and dyeing preparation processes. The chapter reviews the use of acid and reactive dyes, as well as dyeing silk with direct colours and natural dyes.

Key words: dyeing of silk, degumming, acid dyes, reactive dyes, direct colours, natural dyes.

5.1 Introduction

This chapter reviews the following aspects of dyeing:

- types of dye,
- factors affecting dyeing,
- degumming processes to prepare silk for dyeing,
- bleaching,
- dyeing of silk with acid and reactive dyes, and
- dyeing of silk with direct colours and natural dyes.

5.2 Types of dye used for silk

There are numerous ranges of dyestuff available for use in silk dyeing. Almost every class of dyestuff used for cotton or wool can be used for dyeing silk. In general, the dyestuffs are applied by techniques similar to those used for wool or cotton. The main objective of the coloration of a textile fibre is that the colour should be permanent and should not damage the nature of the fibre. It is essential therefore that the dyestuff should retain its colour during the useful life of the fabric.

The chemical composition of a fibre determines, in a large measure, the type of dye, which is most suitable to apply. The ease with which dyes penetrate a particular fibre structure is also dependent upon the physical structure of the fibre. In silk, there exist regions of differing degrees of molecular order and disorder, i.e., crystalline and amorphous (or non-crystalline) regions respectively (Flensberg and Hammers, 1988). The groups which are present in silks in significant amounts and ionize at an appropriate pH are –COOH and $–NH_2$ (Peters, 1975). The dye–fibre interaction occurs through ionic or covalent bonding. The mechanism of dyeing silk is dependent on free amino and carboxyl groups and also on phenolic with accessible –OH groups. Because of the slightly cationic character of silk with an isoelectric point above pH 5.0, a wide range of dyes can be used, including acid, basic, direct, metal-complex, mordant, natural and reactive dyes (Work, 1976; Shukla and Mathur, 1995). Amongst these, acid and metal-complex dyes are the most widely used. However, reactive dyes are now increasingly used because of their excellent wet fastness which enables machine washing. The application of vat colours is restricted due to the very high alkaline pH of dyebaths which may damage the silk if the temperature of the bath rises.

The choice of colour depends mainly on the grades of shade, brightness and fastness (Gulrajani and Gupta, 1989). Acid and metal-complex dyes possess a better affinity for fibre, are easily absorbed, but have poor to moderate washing fastness. Reactive dyes offer good wash fastness and a full range of dischargeable colours with high perspiration fastness. Basic dyed silk has very poor light fastness due to the formation of coloured cations which do not permit reaction with protonized amino groups when dyed in an acidic bath.

5.3 Factors affecting dyeing behaviour

Obtaining uniformity of colour in the dyeing of textile fibres and fabrics is important. Any process, whether chemical or mechanical, which modifies the dyeing properties, that is the dyeability, of the textile substrate must be strictly controlled to avoid irregularities in subsequent dyeing. The diffusion and absorption of dye depend largely on the chemistry of the molecules, the physics of the fibre structure and its ability to undergo transition before or during dyeing (Burdett, 1975).

It is well known that silk fibres produced in different parts of the world differ not only in their chemical architecture, but also in their morphological properties. The dye–fibre interaction will therefore also differ between varieties. A silk filament obtained from a single cocoon will also exhibit noticeable changes along its length, for instance in filament size, crystallinity and orientation (Iizuka, 1985; Sen and Babu, 1999, 2000). Variation in the microstructure is likely to result in differences in dye pick-up along the filament length.

Any attempt to correlate fibre structure with dyeing behaviour must take into account the two main structural features of fibres which govern their reactivity or dyeability:

1. permeability, or the ease with which dye molecules diffuse into the fibre matrix;

2. the presence of reactive functional groups in the molecular chains of the fibre.

These two features are largely determined during the manufacture or growth of the fibre and the nature of the dye is equally important. Therefore any correlation should take into account:

1. the volume fraction, size, shape and configuration of the regions or voids accessible to the dye;
2. the volume fraction (or degree of crystallinity), size, configuration and distribution of ordered regions or units of the fibre;
3. the type, concentration, distribution and degree of ionization of ioinizable groups in the fibre and the dye;
4. the molecular interaction between all molecular species present.

In the above list, the characteristics of the non-crystalline regions of a fibre are described by item (1), whilst the characteristics of the crystalline regions are described by (2). The parameters of the actual dyeing process, particularly pH, ionic strength and temperature also exert an effect. However, the aqueous environment in which most dyeing processes are carried out is of greater importance. The degree of swelling exhibited by silk fibres when immersed in aqueous solutions is a reflection of the chemical composition of the fibre and may play a significant role during dyeing. Indeed, the degree of swelling in water and the moisture regain of silk may be closely related to its dyeability.

Early work on the mechanism of acid dye uptake in protein fibres gave rise to the hypothesis that under acidic conditions, cationic hydrogen ions, being small, diffuse readily and are first absorbed by the swollen fibre, associating with the amino groups in the fibre to give cationic ammonium groups, $-NH_3^+$. The dye–fibre interactions in the adsorption of anionic dyes by silk are due to electrostatic attraction to positive basic groups in the fibre and to Van der Waals' forces. As the amount of basic amino residues in *Bombyx mori* silk is 0.18×10^{-3} eq/g, the dyeing behaviour of silk with acid dyes is thought to be closer to that of nylon (Mitsubishi *et al.*, 1985).

Ammonium groups act as dye sites for anionic dyes. The maximum possible concentration of ammonium groups, which is equal to the concentration of amino groups initially present (in dry fibre), may also represent the upper limit of dye anions sorbed in achieving saturation. However, the ionic groups do not limit the interaction between the dye and the fibre. In addition to ionic interactions, some level of hydrophobic dye–fibre bonding can be expected, particularly in sparingly soluble dyes with large molecules. The mechanism lies somewhere between the idealized electrostatic attraction (Langmuir isotherms) and the idealized hydrophobic bonding of dispersed dyes in polyester (Nernst isotherms). Both these situations may exist to differing extents within the same fibres.

The interaction of basic dyes with silk fibre involves the dyes ionizing into cations and colourless anions. In these dyes, the coloured portion is derived from the free or substituted amino groups, such as, $-NH_2$, $-N(CH_3)_2$, $-N(C_2H_5)_2$. Basic dyes, being cationic in nature, interact with the carboxylic groups of the protein fibres. When these carboxyl groups are ionized, negative sites are created in the fibre. The cations of basic dyes adhere these sites by electrostatic attraction. For instance,

$$NH_2\text{–silk–}COO^- + \text{Dye} + Cl \rightarrow NH_2\text{–silk–}COO^- + \text{Dye} + HCl$$

Metal-complex dyes are pre-metallized and are applied to silk in a similar manner to the acid dyes. These dyes normally have one or two sulphonyl groups which can combine with the protonated groups of silk by electrostatic forces. Co-ordinate bonds may also be formed between the chromium atom of the dye and suitable groups (hydroxy, amino, imino, etc.) of the fibre. A study of the interaction between direct dye and mulberry silk (Shakra *et al.*, 1999) indicated that the magnitude of dye adsorption by silk depends on:

(a) the presence of sulphonate groups in the dye structure;
(b) the electrical force between dye and silk;
(c) the hydrogen bonding and
(d) Van der Waals' forces.

However, there is no consolidated and well-documented literature which relates specifically to silk–dye interaction and morphological structure. Interestingly, some literature on the effects of dyeing on structure is available. Somashekarappa *et al.* (1998) used wide-angle X-ray scattering (WAXS) to study the changes in the crystal size of mulberry (bivoltine and multivoltine) silk fibres after dyeing with acid and metal-complex dyes.

The crystal size values reduced from 29.7 Å to 20.3 Å in acid dyed fibres and from 29.7 Å to 21.5 Å in metal complex dyed bivoltine silk fibres. In multivoltine varieties, the values reduced from 18.6 Å to 12.7 Å in acid dyed fibres. No significant differences were observed for metal-complex dyed fibres. Tsukada *et al.* (1996) investigated the dyeing behaviour of Japanese mulberry silk fibres. The results of this study indicated the occurrence of large colour differences between fine and coarse silk fibres, regardless of the dye class and dyeing conditions. The dye uptake of fine fibres was lower and was attributed to their higher crystallinity and molecular orientation. However, the authors believe that thorough investigation is needed to establish the relation between silk fibre structure and dyeability. In another study on the effect of grafting on *B. mori* silk fibres, Tsukada *et al.* (1992) reported that dyeability increased after the silk was grafted with ethoxyethylmethocrylate polymer. This was attributed to modifications of the morphology of the amorphous region of the fibre due to the steric interaction between ETMA polymer and the fibroin molecules.

Some attempts have been made to study the dyeing behaviour of mulberry silk using low temperature with ultrasonic techniques using the redox system. A study on dyeing mulberry silk (Saligram and Shukla, 1993) using ultrasound (frequency 20 MHz, output 120 W), showed the dye uptake to be increased by 55–95% at a temperature of 35–45°C over shorter time periods (15–30 min). The increase in dye uptake was attributed to the breaking up of dye aggregates, as also due to removal of air from the fibre capillaries thus accelerating the diffusion of the dye (Bryant, 1948). Similar results were obtained on mulberry silk fabrics dyed with acid, basic and metal-complex dyes (Mishra and Venkidusamy, 1992). Dyeing mulberry silk at a low temperature using hydrogen peroxide/glucose by the redox system has been reported by Luo (1991). The results of this study showed that the exhaustion and fixation of acid dyes improved due to an increase in the number of dye sites caused by the action of free radicals. The improvement achieved by using the redox system was attributed to a decrease in the pH of the dye solution due to the decomposition of glucose with hydrogen peroxide (Park *et al.*, 1994).

The most suitable type of dye is largely determined by the chemical composition of the fibre. The ease with which dyes penetrate a particular fibre structure is also dependent upon the physical structure of the fibre. Differing degrees of molecular order and disorder exist within the crystalline and amorphous regions of silk. The crystalline regions provide strength and rigidity, whilst the amorphous regions provide flexibility and reactivity. These are inherent characteristics which are unalterable for all practical purposes.

The diffusion of any substance into a polymer is related to the compactness of its structure. Because dye diffuses through the amorphous regions, changes in the dyeing rate are likely to arise due to differences in the density of these regions. Heat treatments, especially steaming, also modify the structure and properties of fibres (Hiroyuki, 1999). The fibre density, and hence the crystallinity, increases with a rise in temperature. Dye sorption has been reported to reduce after steaming. The curve of dye sorption plotted against the steaming temperature shows a minimum.

In order to correlate silk structure with its dyeing behaviour, the following approaches were examined:

1. Studies on the amino acids of different varieties of silks and their correlation with dyeing behaviour.
2. As the structural parameters, *viz.* crystallinity and orientation vary along the filament length, the different layers of a cocoon were dyed to establish correlation between dye uptake, rate of dyeing, crystallinity and orientation.

3. Studies on the dyeing behaviour of three important non-mulberry silks *viz*. tasar, muga and eri, for comparison purposes with mulberry silks and to explore their commercial importance.
4. Correlation of the dyeing behaviour with changes in the physical structure of silks brought about by heat/steaming treatment.

Silk exhibits higher crystallinity and orientation than wool. Though some work has been reported on the effect of chemical structure on its dyeing behaviour, very little has been reported on the importance of physical structure and its effect on dyeing. Most of the studies reported are on cultivated, that is mulberry silk, and non-mulberry silks have been ignored. In this chapter, details of dyeing silk with various classes of dyes are presented with an overview of the preparatory processes employed, such as degumming and bleaching.

5.4 Preparation of silk for dyeing: degumming

Fibres contain various natural substances. These are pectin, grease, wax, colour tone, inorganic substances, etc. There are secondary impurities such as oil agents and warp starch which facilitate weaving and twisting. However, these impurities obstruct dyeing and affect the fibre, so diminishing the quality of the yarn. It is therefore necessary to remove them before chemical processing. The raw silk thread obtained from a cocoon contains 20–30% sericin protein. Other substances and secondary adhesives are also present, making the silk coarse and lacking in lustre. The composition of the cultivated variety of silk, *B. mori*, is given in Table 5.1.

The process of eliminating sericin, known as 'gum' from raw silk is known as degumming. The purpose of boiling-off silk is threefold:

- to remove the sericin content;
- to remove reagents added by the throwster in soaking raw silk and
- to remove any dirt picked up during reeling, throwing or knitting.

Since all the natural and acquired impurities except sericin, constitute only a very small fraction and are comparatively easily removed, the degumming process may be considered primarily as the process of cleavage of peptide bonds of

Table 5.1 Composition of raw silk

Fibroin	70–80%
Sericin	20–30%
Waxy matter	0.4–0.8%
Moisture	10–11%
Carbohydrates and starches	1.2–1.6%
Inorganic matter	0.7%
Pigment	0.2%

sericin either by hydrolytic or enzymatic methods and its subsequent removal from fibroin by solubilization or dispersion in water (Komatsu, 1985).

Natural and acquired impurities, with the exception of sericin, are present in very small amounts and are comparatively easily removed. The degumming process may therefore be considered primarily as a process of the cleavage of peptide bonds of sericin, by either hydrolytic or enzymatic methods, and its subsequent removal from fibroin by solubilization or dispersion in water. The removal of sericin results in a silk yarn or fabric with soft handle and enhanced lustre.

Degumming mainly involves the removal of sericin from the fibroin. It is synonymous with the scouring process used for the purification of cotton and wool. Sericin is insoluble in cold water but is comparatively easily hydrolysed, which breaks down the long protein molecules into smaller fragments easily dispersed or solubilized in hot water. The hydrolysis of proteins can be carried out by treatment with acids, alkalis and enzymes. Acids are non-specific and tend to attack vigorously. Alkalis also attack both sericin and fibroin. However, the variation in the rate of hydrolysis is large enough for the reaction to be controllable. Degumming with soaps in the presence of mild alkalis such as soda ash is also used. Degumming with alkalis is a function of pH, temperature and duration of the treatment. The pH should be kept at a level of 9.5–10.5. If the level is below 9.5, the process of removing sericin will be slow. If the pH is over 10.5, the degumming loss will be greatly increased.

The different ways of degumming silk are as follows:

- extraction with water;
- boil-off in soap;
- degumming with alkalis;
- enzymatic degumming methods;
- degumming in acidic solutions and
- degumming with organic amines.

The dissolvability of sericin poses a problem in degumming. Sericin is a protein similar to fibroin but its behaviour in water is more like that of gelatin. Its basic character changes due to the action of heat, acid, alkali and heavy metallic ions. Sericin dissolves easily in hot water, especially if the water contains soap or alkali. The difference in the rate of hydrolysis of sericin and fibroin is large enough to permit control of the reaction.

The degumming of silk with enzymes is based on the specificity of the attack of enzymes on certain amino acid adducts. Because sericin and fibroin contain different proportions of some important amino acids, the enzymes attack them to different extents. Some studies (Gulrajani *et al.*, 1992; Gulrajani and Chatterjee, 1992) have shown the possibility of degumming silk by the use of acids. It appears that the degumming action of acids is due to hydrolysis of the proteins of some amino acid residues which are found in greater proportions in sericin than in fibroin. Attempts have also been made to degum silk using amines (Gulrajani and Malik, 1993; Gulrajani and Sinha, 1993). This is an entirely new concept and conditions have been optimized for the use of certain amines as degumming agents.

5.4.1 Extraction with water

Silk will not dissolve in water at room temperature but is highly susceptible to dissolution in boiling water. Extraction with water under pressure at 115°C has been used by various investigators (Sheldon and Johnson, 1925; Mosher, 1932a) to isolate sericin from fibroin and to classify it into different fractions. In order to remove sericin from raw silk, the yarns must be autoclaved for long periods with water at temperatures over 100°C. It has been suggested (Dunn *et al.,* 1944) that for the complete removal of sericin in cultivated varieties of silk it is necessary to extract the silk yarn with water at 120°C over four periods of 2 h.

The sericin in fresh cocoons has a non-crystalline molecular structure and is almost 100% water soluble protein. Due to conditions inside the cocoon, storage conditions and the processes of yarn making, it changes or is denatured and its solubility undergoes a marked change. During the reeling process, silk cocoons are kept in a hot water bath at 50–60°C and as a result, about 1–3% of the sericin is dissolved in the water (Carboni, 1952). Subsequent heating of raw silk yarn in water at high temperatures over varying time periods results in the gradual removal of sericin (Gulrajani *et al.,* 1992).

5.4.2 Boil-off in soap solution

Degumming silk by boiling-off in soap solutions has been practised for more than 200 years. The recommended standard method uses Marseilles soap, which is obtained from olive oil. Many other soaps have been tried with some degree of success. These include palm-oil soap, lard-oil soap and oleic acid soap.

Tsunokaye (1932) has postulated that the degumming action of soap solution is due to alkali formed by hydrolysis of the soap. The alkali forms a chemical bond with sericin and produces soda salt. The swollen sericin is separated by the soap and dissolves in water due to the emulsification action of the soap. The quantity of soap required for complete degumming depends upon the nature and type of silk. Generally, 20–30% soap to the weight of material is used, which is 6–7 g/L for a liquor ratio of 30–40. A degumming time of 90–120 min at boil is considered sufficient. Soaps with the greatest capacity to hydrolyse are most effective in degumming. Olive-oil soap is considered particularly suitable because of its high degree of hydrolysis which gives better lustre.

Because soap is relatively mild in action and its permeability is high, there is less likelihood of over degumming. It also enhances the whiteness, lustre and feel of the fabric and has some softening effect. It has been observed that silk degummed at 90°C has a superior colour and feel when compared to that treated at boiling temperature (Mosher, 1932). Soaps made from unsaturated fatty acids or oils leave a better lustre on the silk than those produced from saturated fatty acids.

However, there are certain disadvantages to degumming with soaps. The quality of the water used with the soap will influence the silk quality. Metallic ions in the water, such as calcium and magnesium, combine with the soap and produce an insoluble metallic soap which deposits on the fibre and reduces the lustre. The residual soap in the silk causes problems such as uneven dyeing, yellowness and embrittlement. The quantity of soap required is quite high (about 20–30% on the weight of fibre) and, when subsequently discharged, is a cause of pollution. The cost of imported Marseilles soap is high; therefore degumming is usually carried out using non-standard indigenous soaps based on sodium stearate.

Attempts have been made to overcome some of these disadvantages by using a mixture of soap and alkali (Brezezinski and Malinowska, 1989). This combination accelerates the process of degumming and is superior on grounds of both economy and quality, but has been found to be susceptible to hard water. Some of the alkalis recommended for addition to soap solution are: sodium silicate (Morgan and Seyferth, 1940; Brezezinski and Malinowska, 1989), sodium sesquicarbonate (Morgan and Seyferth, 1940), sodium carbonate (Brezezinski and Malinowska, 1989), sodium phosphate, sodium hydroxide and a buffer mixture of sodium carbonate and sodium bisulphite (Markuze and Maleev, 1941). Leggis (1938) has reported that high-grade olive-oil soap liquor gives a starting pH of 9.2–9.5 which rapidly drops to 8.5; the addition of alkali tends to maintain pH at the effective level.

5.4.3 Degumming with alkalis

Alkalis have a strongly destructive effect on proteins. The severity of treatment required to remove sericin from a given sample depends on the type of silk. Degumming with soaps in the presence of alkalis is a long-standing practice. Here soap acts as the degumming agent and the alkalis aid the process. Alkali based methods reduce costs through improving productivity and machine efficiency by cutting the cost of chemicals and reducing the treatment time from 90–120 to 10–30 min. Although soaps are superior to alkalis due to their surface active properties, the cost involved is higher. Alkalis hydrolyse proteins by attacking the peptide bonds. The amino acids reported to be mainly affected are cysteine, serine, threonine and arginine (Sanger, 1952). The process has to be carried out under controlled conditions to avoid over degumming. Generally, the more alkaline a solution is, the higher will be the dissolution power.

It has been established by Mosher (1930c) that the pH of a boil-off bath in alkaline degumming should be maintained between 9.5 and 10.5. Below a pH of 9.5, the rate of degumming is too slow, causing prolonged exposure to mechanical damage; above 10.5 the danger of chemical damage increases rapidly. Degumming is incomplete and very slow at pH below 8.4 (Scott, 1925). In a review (Gulrajani, 1992) investigated various alkalis for use as degumming agents. These are: sodium hydroxide, sodium carbonate, potassium carbonate, sodium silicate, trisodium phosphate, sodium pyrophosphate, sodium bicarbonate, borax and ammonia. Sodium carbonate is the preferred alkali.

The use of buffer systems rather than individual alkalis is recommended to maintain the desired pH as they are considered to be safer. Some of the buffers investigated are: potassium carbonate–sodium bicarbonate (Bryant, 1950), sodium carbonate–sodium bicarbonate (Gulrajani *et al.*, 1990), disodium hydrogen phosphate–trisodium phosphate (Gulrajani *et al.*, 1990) and potassium tetraborate–boric acid. Sodium carbonate–sodium bicarbonate is the most widely investigated buffer. In a study on degumming Murshidabad silk, Gulrajani *et al.* (1990) found that the optimum pH for degumming is 10.2, obtained by using 0.1 molar solution each of sodium carbonate and sodium bicarbonate. The optimum treatment conditions were worked out to be as follows.

Degumming with alkali alone results in silk with poor wetting properties. However if a non-ionic surface active agent is introduced along with the alkalis, the wetting properties improve considerably without affecting the rate of degumming (Gulrajani *et al.*, 1990). It has also been reported (Sadov *et al.*, 1978) that the rate of degumming depends not only on the pH but also on the total concentration of electrolytes in the buffer solution. For instance, 20 min are required for silk degumming at a temperature of 90°C in a 0.1 N carbonate–bicarbonate solution with a pH = 10.00; 30 min are required to degum silk in a 0.05 N solution and 60 min are required in a 0.02 N solution. The decomposition of fibroin also increases with an increase in normality of the carbonate–bicarbonate solution at the same pH.

The presence of salts in the water enhances the degradation of silk treated with alkalis, due to the distribution of ions in the external and internal phases, according to the Donnan membrane theory. The addition of salt to an alkaline solution in which negatively charged fibroin is present increases the concentration of –OH groups in the internal phase, which in turn attacks the fibroin. The addition of metal-chelating agents such as $Na_3HP_2O_7.H_2O$ (sodium hydrogen pyrophosphate) prevents damage to the silk during degumming with alkalis (Obo, 1979).

5.4.4 Degumming in acidic solutions

Some acids such as sulphuric, hydrochloric, tartaric and citric acids can be used as degumming agents. Viktorov and Bloch (1933) carried out studies on the degumming of silk with sulphuric and hydrochloric acids. The action of organic acids was reported to be much less pronounced than that of mineral acids. Flensberg and Hammers (1988) have reported that acid degumming is being carried out on a commercial scale. These investigators degummed silk with tartaric and citric acid and observed that in terms of tensile strength and elongation of the silk fibres, these treatments gave very good results, some of which are comparable with those obtained by conventional degumming while some are superior.

Studies in the degumming of silk yarn and fabric with ten organic acids have been carried out (Gulrajani *et al.*, 1992) using the following conditions:

- Concentration: 0.05 mole/L. Treatment time: 60 min. Temperature: 100°C. Non-ionic surfactant: 3 g/L.

5.4.5 Degumming with enzymes

Enzymes are proteins which catalyse a specific chemical reaction and are known as 'biocatalysts'. Generally, enzymes appear in living cells and therefore work at atmospheric pressure and in mild conditions (e.g. 40°C, pH 8.0). Different enzymes may cause hydrolysis, reduction, oxidation, coagulation and decomposition reactions. Hydrolytic enzymes, for example cellulose, trypsin and papain, are not commonly used in the textile industry. However trypsin, papain and bacterial enzymes are the main types of enzymes used in silk degumming. These enzymes are known as 'proteases' because they degrade proteins, producing polypeptides, peptides and other substances by hydrolysis of the –CO–NH– linkage. The properties of any type of protease should therefore be studied before use. Sericin is mostly easily removed when it becomes alkaline and so the proteases which show stability and activation in an alkaline bath are considered to be the most effective. Trypsin and papain are most effective in hydrolysing peptide bonds formed by the carboxyl groups of lysine and arginine. Sericin and fibroin contain these amino acids in different proportions.

A patent (Japan Kokai, 1991) has claimed that treatment with a liquid containing 4% (owf – on weight of fabric) 2:2:1 (weight ratio) $NaHCO_3$: Na_2CO_3: Na_2SO_3 mixture and 0.3% alkali protease at 50–60°C for 10–20 min gives good uniform results as compared to those obtained with a liquid containing only Na_2CO_3 and protease. A pre-treatment with weak alkaline solutions of sodium silicate (Boneva *et al.*, 1990) or sodium carbonate or borax followed by degumming with pancreatin enzyme (trypsin) has been claimed to give improved results.

Trypsin

Trypsin is a serine protease which is secreted by the pancreas and is most active in the pH range between 7 and 9 at 37°C. It reacts with peptide bonds between the carboxylic acid group of lysine or arginine and the amino group of the adjacent amino acid residue. Ammonium bicarbonate (0.1 M) is considered to be a good buffer for this process. An enzyme concentration of 1% or 2% at 37°C for 1–4 h is considered optimum for tryptic digestion.

A detailed investigation of the effect of trypsin concentration, temperature and time of treatment on the removal of sericin from silk by hydrolysis has been carried out by Krysteva *et al.* (1981). According to these investigators, 1% solution of trypsin completely hydrolysed sericin in 10 h at 37°C while the amount of sericin removed in 4 h by treatment with 1% and 8% trypsin solution were 26.4% and 28.7%, respectively. The tensile strength of silk fibres conventionally treated with sodium carbonate was less than that of trypsin-treated fibres.

Papain

Papain is a sulphydryl enzyme isolated from papyra latex and is readily available from a number of sources. It is optimally active between pH 5 and 7.5 at 70–90°C and requires activation by a sulphydryl reagent. Activated papain attacks the peptide bonds between the carboxylic acid group of lysine or arginine and the adjacent amino acid residue. A small cleavage occurs at the carboxylic acid group of histidine and also of glycine, glutamic acid, glutamine, leucine tyrosine residues. Papain is the only plant protease which has been extensively investigated for degumming of silk.

Degumming silk with papain requires pre-treatment with hot water to swell the sericin. Subsequent incubation of the enzyme is carried out in the degumming liquor under suitable conditions of pH and temperature in the presence of an activator. Originally, the use of papain at 50–70°C in the presence of potassium cyanide or hydrogen sulphide was suggested. Later, other activators of a less poisonous character. For example, sodium hydrosulphide, sulphoxylate or sulphite were also recommended. The process of degumming with papain and the properties of the degummed silk substrates studied by various investigators has been recently reviewed (Gulrajani, 1992).

Bacterial enzymes

Degumming can be achieved by treatment with bacteria. A bacterial enzyme Alcalase, marketed by NOVO, has been found to be very effective in hydrolysing sericin. A study by Lee *et al.* (1986) found this enzyme to be more effective than trypsin and papain. It can completely hydrolyse sericin in one hour at 60°C and pH 9. The physico-mechanical properties of Alcalase-treated fabrics have been found to be better than those of conventional soap-soda boiled silk. Another alkali stable protease found suitable for degumming of silk is Degummase which is marketed by Advanced Biochemicals, India. It has been found to be most effective at 50°C and a pH of 8.75. These studies have shown enzymes to be less mild and safe than was formerly thought. The use of enzymes at higher concentrations and for longer periods of time may lead to degradation of the silk.

Degumming with organic amines

Amines ionize in water to give hydroxyl (OH) ions which can act as degumming agents for silk. The action of amines thus would be similar to that of alkali:

$$RNH_2 + H_2O \rightarrow RNH_3^+ + OH^-$$

Attempts have been made to carry out degumming by using amines (Gulrajani and Malik, 1993; Gulrajani and Sinha, 1993). Degumming with organic amines can be carried out with 0.2 mole/L at 80°C for 30 min using methylamine, ethylamine, diethylamine, triethylamine or ethylene diamine. However, as triethylamine and ethylene diamine are the most effective as they have a fairly high boiling point.

5.5 Bleaching of silk

The fibres spun by silkworms contains natural colouring matter tinted with yellow, yellow-green and brown pigments. These are associated mainly with sericin and hence are eliminated during degumming. However, residual pigments are adsorbed by fibroin and therefore silk fabrics made from yellow raw silk are cream coloured after degumming. The removal of sericin from the silk results in material which is dull white or lightly tinted. As some of the sericin is closely held by fibroin, complete elimination of colour by degumming is not possible. During the bleaching process, these natural colouring matters are de-colourized or removed to produce pure white material. An efficient bleaching process must ensure pure and permanent whiteness, even dyeing properties and must not degrade the material.

Silk is bleached by the use of either reducing agents or oxidizing agents. However, material bleached with reducing agents tends to re-oxidize and the original colour may be restored. Oxidizing bleaching is therefore preferred. The most commonly used reducing agents are sulphur dioxide, sodium hydrosulphite and sodium or zinc sulphoxylate formaldehyde. The oxidizing agents used are potassium permanganate, sodium perborate, sodium peroxide or hydrogen peroxide, the latter being the preferred bleaching agent. Chlorine-based bleaching agents are bleaching powder, sodium hypochlorite and sodium chlorite. These are not widely used as they tend to chlorinate the fibroin. Various methods of bleaching are summarized below.

Bleaching with sodium hydrosulphite (hydros)

(a) The degummed silk is treated in a bath containing 4 g/L of sodium hydrosulphite at 50°C for 4–6 h.
(b) The material must be completely immersed and the bleaching liquid stirred sufficiently to ensure uniform distribution.
(c) Finally the material is washed thoroughly. The solution should be made up only when required as its reducing power deteriorates rapidly under storage.

Bleaching with sulphur dioxide

(a) Degummed silk goods are hung in a suitable chamber where they are exposed to sulphur dioxide gas or immersed in sodium bisulfite in water for 4–6 h.
(b) Approximately 5 kg of sulphur is required to bleach 100 kg of silk. This method of bleaching is not generally practised as it causes air pollution.

Bleaching with sulphoxylate

(a) Sodium sulphoxylate formaldehyde is a stable reducing agent marketed as rongolite, sofolite, etc.
(b) A bath containing 2–4% (owm – on the weight of material) sodium sulphoxylate formaldehyde and 1–2.5% (owm), formic acid (85%) is prepared.
(c) The goods are treated at boil for 20–30 min. Following this, the material is thoroughly washed.
(d) Zinc sulphoxylate formaldehyde, known as decolin and safolin, can also be used as a bleaching agent.

Bleaching with sodium peroxide

A typical recipe for this process is as follows:
- sodium peroxide – 2–4% (owm)
- magnesium sulphate – 0.5% (owm)
- sodium silicate – 3% (owm)
- sulphuric acid (96%) – 2–4% (owm)
- sodium bicarbonate – 0.5% (owm)

- material:water ratio – 1:30
- time: 4–5 h at 60°C

The oxygen yielded by the decomposition of sodium peroxide oxidizes the chromogens of silk so that it becomes colourless.

Bleaching with hydrogen peroxide

Hydrogen peroxide is the preferred bleaching agent and is sold as a 35–50% aqueous solution. It is quite stable under acidic conditions. A typical recipe for the peroxide bleaching process is as follows:
- silk goods – x kg
- material:water ratio – 1:30
- H_2O_2 – 6 g/L
- sodium silicate – 1.5 g/L
- soda ash – 0.5 g/L
- temperature – 80–85°C
- time – 60–90 min

The liberation of perhydroxy ions acts as the bleaching agent in this process. The sodium silicate acts as a stabilizing agent and aids in maintaining a slow and steady processing rate. The hydrogen peroxide is an important agent in ensuring lasting white colour. It also possesses good stability in storage.

5.6 Dyeing of silk with acid dyes

Acid dyestuffs are so called because the substances were originally applied in a bath containing mineral or organic acid. They are nearly all sodium salts of organic acid in which the anion is the active coloured component. Acid dyes are usually sodium salts of sulphonic acids, less frequently of carboxylic acids, and are therefore anionic in aqueous solution. They will dye fibres with cationic sites and are usually substituted for ammonium ions in wool, silk and nylon. These fibres absorb acids. The acid protonates the fibre's amino groups, making them cationic. Dyeing involves exchange of the anion associated with an ammonium ion in the fibre, with a dye in the bath. The reaction of silk fibre with acid dye is shown in Fig. 5.1. Acid dyes produce brilliant shades with good light fastness but poor to moderate wash fastness which can be improved by treatment with a cationic dye-fixing agent (2–4 g/L) at 40–50°C for 20 min. Acid dyes generally show poor coverage on silk, necessitating topping with basic dyes to produce even shades.

$$\text{Fibre—NH}_2\text{(s)} + \text{H}^+ \text{(aq)} + \text{HSO}_4^- \text{(aq)} \longrightarrow \text{Fibre—NH}_3^+ \text{HSO}_4^- \text{(s)}$$

$$\text{Fibre—NH}_3^+ \text{HSO}_4^- \text{(s)} + \text{Dye—SO}_3^- \text{(aq)} \longrightarrow \text{Fibre—NH}_3^+ \text{Dye—SO}_3^- \text{(s)} + \text{HSO}_4^- \text{(aq)}$$

5.1 Reaction of acid dye with silk fibre.

Table 5.2 The chemical constitutions of some typical acid dyes

Parameters	Levelling dyes	Fast acid dyes	Milling dyes	Super-milling dyes
Acid used	Sulphuric	Acetic	Acetic or NH_4^+	NH_4^+
Dyebath pH	2–4	4–6	5–7	6–7
Migration ability	High	Moderate	Low	Very low
Washing fastness	Poor–fair	Good	Very good	Very good
Molecular weight	Low	Moderate	High	Very high
Dye solubility	High	Moderate	Low	Low
State in solution	Molecular	Aggregated	Aggregated	Aggregated
Substantivity (pH 6)	Very low	Moderate	High	High

5.6.1 Chemical structure of acid dyes

There are many different chemical types of acid dyes (Table 5.2). Sulphonated azo dyes constitute the major group and are mainly mono and bis-azo compounds which range in colour from yellow, through red to violet and brown (Fig. 5.2). There are some navy-blue bis-azo dyes that can be built up to give blacks. Azo dyes for polyamide and protein fibres have greater substantivity. The higher their molecular weight, the lower the number of sulphonate groups per molecule. Anthraquinone acid dyes complement the azo dyes and range in colour from violet through blue to green. These dyes often have very good light fastness. Acid dyes with triphenylmethane (blues and greens) and xanthene (reds and violets) chromophores are less important types which are noted for their brilliant colours and often have poor light fastness. Sulphonated copper phthalocyanine dyes provide bright turquoise dyes of very good light fastness. Acid dyes are commonly classified according to their dyeing behaviour, especially in relation to the dyeing pH, their migration ability during dyeing and their washing fastness.

The molecular weight and degree of sulphonation of the dye molecule determine these dyeing characteristics. The original classification of this type, based on their behaviour in silk dyeing, is as follows:

- level dyeing or equalizing acid dyes;
- fast acid dyes;

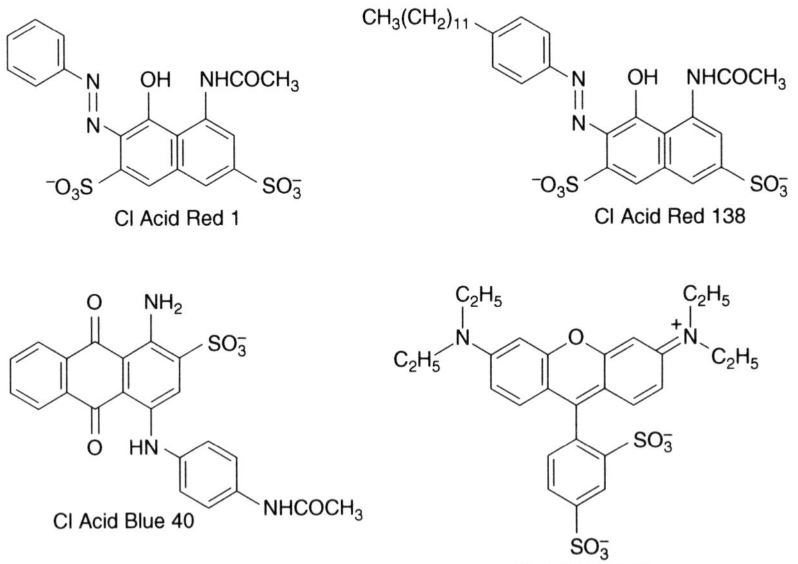

5.2 Chemical structure of some typical acid dyes.

- milling acid dyes and
- super-milling acid dyes.

5.6.2 Acid dyeing process

Exhaust dyeing at neutral or acidic pH is a widely used method which starts at 40–45°C, followed by the addition of CH_3COOH (2–4%) and Na_2SO_4 (5–10%) to the bath to maintain a pH~4–5. Dye is added and the dyeing continued for 10 min. The temperature is raised to 80–85°C over 45 min; dyeing at higher temperatures reduces lustre. Dyeing is continued for a further 30–90 min to promote diffusion of the dye into the fibres, so improving wash and rubbing fastness. A thorough wash then removes superficial dyes. Strong acid dyes can also be applied at pH~3–4 using 1–3% HCOOH (85%) and 5% glauber's salt.

5.7 Dyeing of silk with reactive dyes

The four characteristic features of a typical reactive dye molecule are: a reactive group (R) containing a leaving group (X), a chromophoric group (C), a bridging group (B) and a solubilizing group (S), as shown in Fig. 5.3. Reactive dyes produce bright shades with good wash, light and perspiration fastness due to their reaction with $-NH_2$ groups of silk which forms a covalent bond between reactive groups of dye and nucleophiles on the fibre

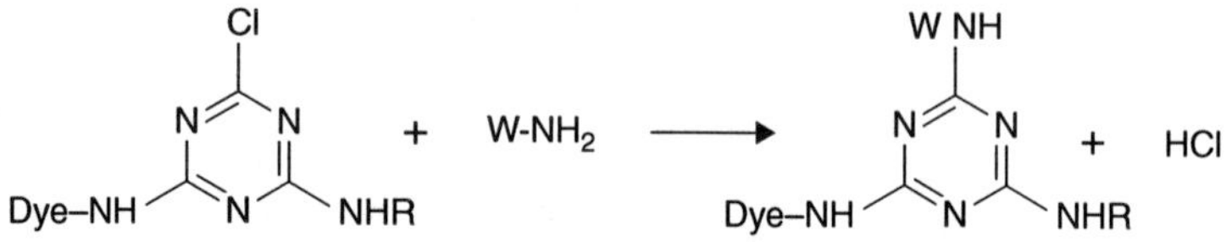

5.3 Typical chemical structure of a reactive dye.

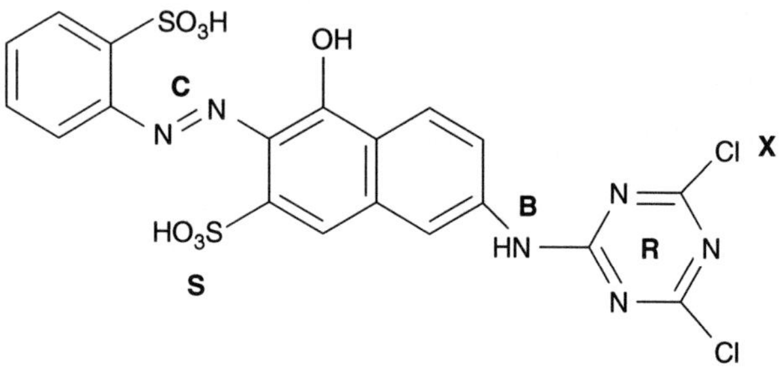

5.4 Reaction of a mono-chlorotriazine dye with the amino group of a protein fibre.

(Gulrajani, 1993). Thorough degumming is a prerequisite for even dying as a similar reaction takes place with any sericin present, causing a loss of colour value. Weighted silk can also be dyed but the build-up is poor.

Fibre reactive dyes are anionic, water-soluble, coloured organic compounds. They are capable of forming a covalent bond between reactive groups of the dye molecule and neucleophilic groups on the polymer chain within the fibre. Consequently, the dyes become chemically part of the fibre by producing dye–polymer linkages and covalent dye–polymer bonds are formed, for instance with the hydroxyl groups of cellulose, the amino, hydroxyl and mercapto groups of proteins and the amino groups of polyamides (Fig. 5.4).

The possibility of forming a covalent bond between dyes and fibres had long been attractive to dye chemists as attachment by physical adsorption and mechanical retention has the disadvantages of low wash fastness or high cost. It was anticipated that the covalent attachment of dye molecules to the fibre would produce very high wash fastness because covalent bonds are the strongest known binding forces between molecules. The energy required to break this bond would be of the same order as that required to break covalent bonds in the fibre itself. Reactive dyes were initially introduced commercially for application to cellulosic fibres and this is still their most important use. Reactive dyes have also been developed commercially for application on protein and polyamide fibres.

Theoretically, all reactive dyes can be used for silk dyeing. However, to achieve the best quality, reactive dyes must satisfy the following requirements (Uddin and Hossain, 2010):

- Brilliancy of shade: This is especially important for silk because many dyes produce a much duller effect and the dyed silk shows a lower degree of exhaustion.
- pH of the dye bath: Silk is damaged in an alkaline medium at high temperatures, so the reactive dyeing should be carried out in an acidic or neutral dyeing bath.
- Good storage stability: The consumption of dyes is small, so the dyes should be highly stable in storage.
- Degree of covalent bonding: A high degree of covalent bonding should be achieved at the end of the dyeing process, so minimizing the clearing treatment required to give maximum wet fastness.
- The rate of adsorption should be higher than the rate of reaction otherwise the dyeing will be uneven.
- The reactivity should be moderate. A highly reactive dye will react even at low temperature reducing the possibility of levelling and migration. A dye with low reactivity requires a longer period at boil with consequent damage of the material. There are currently three ranges of reactive dyes commercially available for use on silk (Table 5.3).

Table 5.3 Commercial reactive dyes for silk

Commercial name	Reactive group	Year of introduction
Lanasol (CGY)	$-NHCO-C(Br)=CH_2$ α-bromoacrylamido	1966
Drimalan F (S)	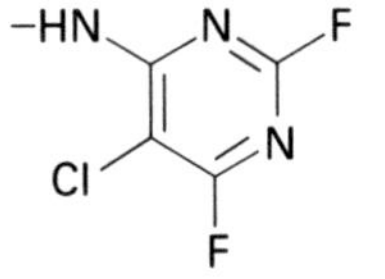 –HN– (pyrimidine ring with N, F, N, F, Cl substituents) 5-chloro-2,4-difluoropyrimidyl	1969
Hostalan (HOE), Hostalan E (HOE)	$-SO_2CH_2CH_2-N(CH_3)-CH_2CH_2SO_3H$ *N*-methyltaurine-ethyl sulphone $-SO_2CH_2CH_2OSO_3H$ β-sulphatoethyl sulphone	1971

$$D-NH-C(=O)-CH(Br)-CH_2Br \xrightarrow{H_2O} D-NH-C(=O)-C(Br)=CH_2 + HBr$$

5.5 Elimination of hydrogen bromide.

Lanasol dyes (Ciba, 1966) are most successful reactive dyes for use on silk. Trichromatic dyeing (i.e. matching a large number of shades with three dyes) is possible using Lanasol Yellow 3G, Blue 3G, and Red 6G (CI Reactive Yellow 39, Blue 69 and Red 84, respectively) as they have almost identical dyeing properties. These dyes are noted for their brightness, high reactivity and good all-round fastness to light and wet treatments. They are essentially bi-functional dyes, provided sufficient neucleophilic groups are available for reaction and these reactions are not sterically hindered. They reported to be α, β-dibromopropionyl amide dyes, which are converted to α-bromoacrylamide by the simple elimination of hydrogen bromide through dissolving in water (Fig. 5.5). Some Lanasol dyes have two α-bromoacrylamide groups, for example Lanasol Red 2G, Lanasol Scarlet 3G, Lanasol orange R (respectively CI Reactive Red 116, Red 178 and Orange 68). The level of fixation of these dyes is particularly high which gives very high wet fastness properties. These dyes may be considered as tetra functional.

5.7.1 Reactive dyeing techniques

Silk dyeing can be divided into three steps in a similar process to that of cellulose dyeing:

1. adsorption of the dye to the fibre surface from the dyeing bath;
2. diffusion of the dye into the interior of the silk fibre;
3. fixation of the dye with reaction centres in fibroin – the following factors have to be considered when choosing the dyeing conditions:
   - the reactivity of the dye;
   - the ratio of fixation and hydrolysis of the dye;
   - protonation of the fibroin and
   - damage to the silk fibre.

Three modes of dyeing are generally used:

1. exhaustion in acidic or neutral bath and fixation in alkaline media;
2. exhaustion and fixation by a one-step process in an acidic or a neutral bath and
3. cold pad-batch in alkaline medium.

The maximum fixation of dichlorotriazine dyes occur in a weak acidic medium at 60°C over a period of 2 h. The dye uptake depends on free –NH_2 and –NH groups of histidine and the fixation can be increased by after-treating with 2% aqueous pyridine or 1% sodium bicarbonate to enhance the reaction of the dye with hydroxyl groups of tyrosine, serine and threonine. It has been reported that silk can be most effectively dyed at 70°C for 1 h at pH ~ 5.6 in presence of Na_2S0_4 (4 g/L). These dyes can also be applied at low temperature in a pad-batch method, in which fabric is padded with dye solution at pH~8 and then treated for five hours at 20–22°C, followed by a cold and hot wash, soaping at 65°C for 15 min with a final cold wash. Impregnating fibres in diethyl phthalate at 90°C for 10 min while heating, followed by treatment with alkali increases the percentage fixation of dyes.

Vinyl-sulphone dyes are marketed as sulphato-ethyl sulphone derivatives and are converted into vinyl-sulphone form by treating with alkali. Fixation occurs by a nucleophilic addition mechanism. The fastness properties are excellent: light fastness is comparable to that of dyed cellulose and the perspiration fastness is good; maximum fixation occurs at pH ~ 7–8. The dye sorption is not dependent on the number of protonated –NH_2 groups on the silk. The addition of electrolytes improves both exhaustion and fixation of dye, while the addition of urea decreases the rate of fixation. Hydrophobic dye (having two methyl groups) has the fastest dyeing rate while hydrophilic dye (having two hydroxyl ethyl groups) has the slowest. N-methyl taurine derivatives show a maximum fixation at pH ~ 6–7 at 95°C for 60–90 min in the presence of NaCl (40–80 g/L).

5.8 Dyeing with direct colours and natural dyes

As with acid dyes, direct dyes are attached to silk by electrostatic bonds between protonated amino groups (NH_2) of the fibre and the dye anion (RSO_3) as well as by hydrogen bonds. Direct dyes produce better fast shades on silk than acid dyes; some direct dyed shades even possess excellent brightness without requiring subsequent after-treatment. The dyeing of silk fabric in light colours is started in a weak alkaline dye bath containing soap (2–3%) and NaCl (5–10%, retarding agent). Dyeing in medium and deep shades is carried out in the presence of Na_2SO_4 (10–20%). The process is started at 40°C, slowly raised to 90–95°C and continued at this temperature for 1 h. The dyed material is then rinsed in warm water and treated with CH_3COOH (5 g/L) at 30°C for 15 min. Direct dyes, which are poorly exhausted at a neutral pH, are dyed in the presence of CH_3COOH (2–5%) and Na_2SO_4 (10–20%). Dyeing begins at 40°C, is raised to 90°C and continued for 45–60 min. The use of ammonium acetate or sulphates in place of acid produces even shades with uniform distribution of the dye.

Natural dyes produce moderately bright colours on mulberry silk, but deep shades can only be produced on tussar, spun and textured varieties. These dyes often require a mordant to make the colour permanent. A few dyes have good light and wash fastness, although some of the mordants are harmful to silk. Colouring matter is extracted by a complex series of processes from the roots, stems, leaves and flowers of various plants, as well as from certain insects and shell-fish. Bright shades are produced with turmeric, berberis, dolu (yellow), annato (orange) and henna (brown). The substantivity of these dyes could be partly due to the presence of tannins which act as a natural mordant. Substances which assist in the dyeing process include acetic acid to neutralize calcareous water, cream of tartar which brightens colours when used in conjunction with mordants, and Na_2SO_4 to control even dyeing. The use of boiled-off liquor causes colouring matter to be attracted more slowly and evenly by the silk, thus helping to preserve lustre.

5.9 References and further reading

Boneva, V., Anastasova, A. and Botev, D. (1990), *Przegl. Wlok*, **44**(1), 21; vide CA., 114, 104129n.

Brezezinski, S. and Malinowska, G. (1989), *Melliand Textilber*, **70**(4), E120, 291.

Bryant, F. (1948), *Text. J. Australia*, **23**, 190.

Bryant, F. (1950), *Text. Res. J.*, **20**, 735.

Burdett, B.C. (1975), *Theory of Colouration*, ed. C.L. Bird and W.S. Boston, Society of Dyers and Colourists Worshipful Company of Dyers, London, England, 111–162.

Carboni, P. (1952), *Silk-Biology Chemistry, Technology*, Chapman & Hall, London, 20.

Dunn, M.S., Carnien, M.N., Rockland, I.S., Shankman, S. and Goldberg, S.C. (1944), *J. Bioi. Chem.*, **156**, 715.

Flensberg, H. and Hammers, J. (1988), *Textil Praxis Internat.*, **43**, 739.

Gulrajani, M.L. (1992), *Rev. Prog., ca.* **22**.

Gulrajani, M.L. and Chatterjee, A. (1992), *Indian J. Fibre. Text. Res.*, **17**, 39.

Gulrajani, M.L., Das, S. and Sethi, S. (1990), *Indian J. Fibre Text. Res.*, **15**, 173. J.P. 01, 229, 804.

Gulrajani, M.L. and Gupta, S. (1989), *Silk Dyeing, Printing and Finishing*, Indian Institute of Technology, Delhi.

Gulrajani, M.L. and Malik, R. (1993), *Indian J. Fibre. Text. Res.*, **18**, 72.

Gulrajani, M.L., Sethi, S. and Gupta, S. (1992), *SDC.*, **108**, 79.

Gulrajani, M.L. and Sinha, S. (1993), *SDC.*, **109**, 256.

Komatsu, K. (1985), *Proc. 7th Int. Wool Text. Res. Conj.*, Tokyo, 1, 373.

Krysteva, M., Arsov, A., Dobrev, I *et al.* (1981), *Int. Conf. Chern. Biotechnol. Biol., Act. Nat. Prod.*, **3**(2), 150; vide CA., 97, 93797f.

Lee, Y.W., Song, K.E. and Chung, J.M. (1986), *Han'guk Chamsa Hakhoechi*, **28**(1), 28–31.

Leggis, W.T. (1938), *Textile World*, **87**, 966.

Luo, J. (1991), *JSDC*, **107**(4), 117–120.

Markuze, K.M. and Maleev, V.I. (1941), *Tekstil Prom.*, (3) 36; vide CA., 38, 2213.
Mishra, S.P. and Venkidusamy, P. (1992), *Textile Dyer and Printer*, **2**(19), 27–29.
Mitsubishi, M. Yagi, T. and Ishiwatari, T. (1985), *Proceedings of the 7th Int. Woll Text. Res. Conf.*, V, Tokyo.
Morgan, A.M. and Seyferth, H. (1940), *Amer. Dyestuff Rep.*, **29**, 616.
Mosher, H.H. (1930a), *Amer. Silk J.*, **49**(7), 53.
Mosher, H.H. (1930b), **49**(8), 54.
Mosher, H.H. (1930c), **49**(9), 59.
Mosher, H.H. (1932a), *Amer. Dyestuff Rep.*, **21**, 341.
Mosher, H.H. (1932b), *Textile World*, **29**, 316.
Oarbre, A. (1986), 'Practical protein chemistry', *A Handbook*, ed. Darbre, A. and Waterfield, D., John Wiley & Sons, New York, 124.
Obo, A.M. (1979), *Sanshi Kenkuu*, **110**, 159; vide CA., 93, 133703y.
Park, J.Y., Kim, T.K. and Lim, Y.J. (1994), *Sen-I-Gakkashi*, **50**(4), 175–179.
Peters, R.H. (1975), *Textile Chemistry*, **III**, Elsevier Scientific Publisher, London, 681–720.
Sadov, F., Korchagin, M. and Matestsky, A. (1978), *Chemical Technology of Fibrous Materials*, Mir Publications, Moscow.
Saligram, A.N. and Shukla, S.R. (1993), *American Dyestuff Reporter*, **8**, 41–43.
Sanger, F. (1952), *Adv. Protein Chem.*, **7**, 2.
Scott, W.N. (1925), *Amer. Dyestuff Rep.*, **14**, 145.
Shakra, S., Allam, E.E. and Mansour, H.F. (1999), *American Dyestuff Reporter*, **88**, 3.
Sheldon, E.M. and Johnson, T.B. (1925), *J. Amer. Chern. Soc.*, **47**, 412.
Shukla, S.R. and Mathur, M. (1995) *JSDC*, **111**, 342–345.
Somashekarappa, S. Nadiger, G.S., Somashekar, T.H., Prabhu, J. and Somashekar, R. (1998), WAXS studies on silk fibres treated with acid (blue) and metal complex (brown) dyes, *Polymer*, **39**(1), 209–213.
Tsukada, M. Freddi, G., Matsumura, M., Shiozaki, H. and Kasai, N. (1992), *J. Appl. Polym. Sci.*, **44**, 799–805.
Tsukada, M. Obo, M. Kato, H. Freddi, G. and Zaneti, F. (1996), *J. Appl. Polym. Sci.*, **60**, 1619–1627.
Tsunokaye, R. (1932), *JSDC*, **48**, 164.
Uddin, K. and Hossain, S. (2010), A comparative study on silk dyeing with acid dye and reactive dye, *Int. J. Eng. Technol.*, **10**(6), 21–26.
Viktorov, P.P. and Bloch, Z.S. (1933), *Text Prom.*, **11**, 43.
Work, W. (1976), *Text. Res. J.*, **46**(7), 485.

6
Developments in the processing and applications of silk

DOI: 10.1533/9781782421580.140

Abstract: This chapter reviews the processing of silk fibroin into formats such as films, mats and hydrogels, silk non-wovens and fluorescent silks. It also discusses the use of silk in apparel and other applications such as medical textiles and composites.

Key words: silk fibroin, silk non-wovens, fluorescent silk, biomedical applications, silk fibre-reinforced composites.

6.1 Introduction

Silk has been around for thousands of years. Originally cultivated in ancient China, Thailand, India and Japan now all have important silk industries. Silk fibres make particularly soft, fine and smooth fabrics which drape well and have a beautiful lustre and sheen. Silk fabric provides comfort in warm weather and warmth during colder months. For these reasons, silk has been used for centuries for high-fashion clothing as well as nightwear and lingerie. In addition to its use in apparel, silk is widely used for bedding items like bed sheets, linens, blankets and pillow covers. It is also used for curtains. Other home furnishing items like cushion covers and table clothes too are popular. Silk is biocompatible, biodegradable and known to have natural healing properties, especially for burn injuries. This is why silk fabric is used as a material for medical textiles. Its biomedical uses have expanded into new areas such as controlled drug delivery (Kundu *et al.*, 2008).

One of the most popular uses of raw silk fabrics is their use for making designer dresses for women. Silk has long been used in traditional clothing such as kimonos in Japan and cheongsam wedding dresses in China. One can also find a huge variety of men's shirts made from silk at men's clothing stores. Silk's breathability and good absorbency makes it comfortable to wear in warm weather and while active. Its low conductivity keeps warm air close to the skin during cold weather. It is often used for blouses. The thinness of the material allows women to tuck a blouse easily into their skirts or wear a jacket on top without having the jacket look bulky. Silk's comfort and appearance has also made it popular for nightwear such as negligees and pyjamas. It is also a component in cold-weather clothing (e.g. for skiing). It appears frequently in accessories like handbags and headbands and scarves. Silk's lustre and beautiful drape also makes it perfect for many furnishing applications. It is used for upholstery, wall coverings, curtains, rugs, cushions and wall hangings. It is commonly used for sheets and bedding. Its popularity has grown for those allergic to dust mites since dust mites are not attracted to silk fabric. Sericin is used in hygroscopic moisture-releasing polyurethane foams for furniture and interior materials, as well as fabric-care compositions,

Silk has many other uses. Excessive transepidermal water loss (TEWL) is one of the causes of dry skin and skin moisturizers have been used to overcome it. Sericin is known as a natural moisturizing factor (NMR). Sericin can naturally saturate into skin. Sericin gel is prepared by using sericin solution with pluronic and carbopol as a stabilizer to prevent water loss from the upper layer of the skin. It forms a moisturizing, semi-occlusive, protective, anti-wrinkle film on the skin surface imparting an immediate, long lasting, smooth, silky feeling (Padamwar *et al.*, 2005). Other uses of sericin include: light and sunscreen compositions, foam-forming aerosol shaving gels, sericin-coated powders for cosmetics, as a dermatitis inhibitor, nail cosmetics, and chewing gums (Gulrajani, 2005).

6.2 Processing of silk fibroin

Silk from the silkworm *B. mori* has been solubilized then reformulated into new materials. Silkworm fibroin can be solubilized by first degumming or removing the sericin using boiling soap solution or boiling dilute sodium bicarbonate solution, followed by immersion of the fibroin in high-concentration salt solutions such as lithium bromide, lithium thiocyanate or calcium nitrate. After solubilization in these aggressive solvents, dialysis into water or buffers can be used to remove the salts or acids, although premature reprecipitation is a problem unless the solutions are kept at low temperature. Resolubilized silkworm cocoon silk and genetically engineered variants of silk have also been spun into fibres (Hudson, 1998). These fibres (Fig. 6.1) do not exhibit the remarkable mechanical properties of the native materials. Electrospun silk has also been generated from silkworm, spider and genetically engineered silks (Jim *et al.*, 2002, 2004). Solubilized fibroin can also be converted, for example, into films, mats, gels, porous matrices and membranes (Matsumoto *et al.*, 2006).

The development of silk nanofibres introduces a new set of potential uses for silk. Silk nanofibres are attractive candidates for biomedical, electrical and textile applications, including tissue-engineered scaffolds, wound

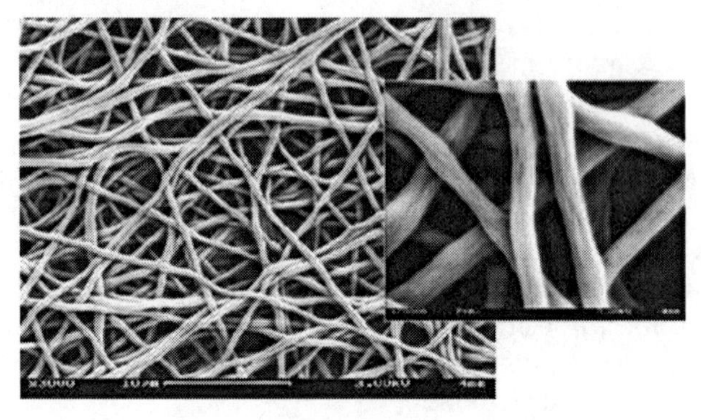

6.1 Images of electrospun fibres from reprocessed fibroin of *B. mori*, illustrating nanoscale fibres with high surface areas combined in a non-woven mat.

dressings and drug delivery systems because of their high specific surface area, increased strength and surface energy and enhanced thermal and electrical conductivity. These fibres can be formed with diameters in the hundreds to thousands of nanometer diameter size range, depending on spinning conditions (solids content, water or organic solvents). These fibres can also be mineralized to further stiffen the materials to expand their potential applications (Li *et al.*, 2005).

6.2.1 Films

Silk fibroin films can be cast from aqueous or organic solvent systems, as well as after blending with other polymers. Films (Fig. 6.2) formed from reprocessed silkworm silk have been produced by air-drying aqueous solutions prepared from the silk solutions after the salts are removed by dialysis. However, rapid gelation can occur at room temperature, so the solutions must be handled carefully. Maintaining solutions of higher concentrations at 4°C significantly slows the gelation process and provides wider processing windows. The films formed from the water-soluble protein generally contain a silk I conformation with a significant content of random coil.

Many different treatments have been used to modify these films to decrease water solubility by conversion of the protein to the silk II polymorph. Most commonly methanol has been used to induce this structural transition. This process was used successfully to entrap enzymes, although the materials embrittle with time (Asakura *et al.*, 1990). These types of silk membranes have also been cast from fibroin solutions and characterized for permeation properties. Oxygen and water vapour transmission rates were

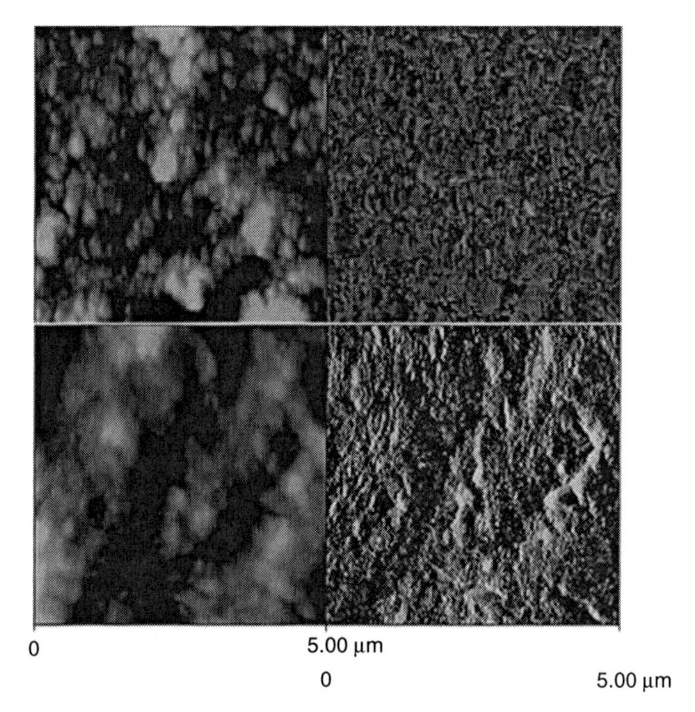

6.2 Atomic force microscopy images of silk films formed from reprocessed fibroin of *B. mori*. Two processing methods are shown: all-aqueous process (top); methanol-induced β-sheet transition (bottom).

dependent on the exposure conditions to methanol to facilitate the conversion to silk II (Asakura *et al.*, 1990). Thin monolayer films have been formed from solubilized silkworm silk using Langmuir techniques to facilitate structural characterization of the protein (Muller *et al.*, 1993).

In other research methanol treatments were avoided and an all-water annealing process for the films was utilized (Muller *et al.*, 1993). This process resulted in a different structural content of the films, reduced silk II content. The films were insoluble in water and retained flexibility over a longer period compared to materials treated with methanol which become brittle rapidly. This approach has been used for films for drug release applications.

Microstructures in silk films are advantageous for increasing surface roughness for cell attachment. Fibroblast attachment to silk films has been shown to be as high as for collagen films. Other mammalian and insect cells have also showed good attachment to silk fibroin films when compared with collagen films. Silk films employed for healing full thickness skin wounds in rats, healed 7 days faster with a lower inflammatory response than traditional porcine-based wound dressings. Chemically modified silk films have also been used for improved cell attachment and bone formation.

6.2.2 Mats

Silk fibroin has been used to generate non-woven silk mats from reprocessed native silk fibres or by electrospinning. Silk fibroin mats 10–30 mm in diameter and pores of about 300 mm in diameter can be obtained. These non-woven mats were studied with a variety of cells, including keratinocytes, fibroblasts, osteoblasts and cell lines from epithelial lung, colon and cervical carcinomas for up to 7 weeks. No degradation of the silk fibres was observed during culture, possibly due to low infiltration of the cells within the matrix.

6.2.3 Hydrogels

Hydrogels are three-dimensional polymer networks which are able to swell in aqueous solutions without dissolving. Hydrogel biomaterials can be used for the delivery of cells and cytokines. Hydrogels are of great interest in medicine, having a variety of applications for drug release, fabrication of prostheses for soft tissues, cells encapsulation and in the field of tissue engineering as matrices for repairing and regenerating tissues and organs. Hydrogels can be formed from reprocessed silkworm silk (Kim *et al.*, 2004). These gels are formed with aqueous solutions of the fibroin prepared as outlined earlier. The rate of sol–gel transition is directly dependent on temperature (the higher the temperature the more rapid the gelation), pH (the lower the pH the more rapid the gelation) and solids content (the higher the solids higher the rates of gelation). Cations can also enhance rates of gelation, with the specific salts dependent on the type of silk. Potassium also plays a role with spider dragline silk and calcium with silkworm silk. The overall rate of gelation has been controlled via osmotic stress, with resulting mechanical, morphological and structural details dependent on the rate and extent of water removal.

6.2.4 Porous 3D sponges

Porous 3D sponges are ideal structures for tissue engineering scaffolds as they provide an environment similar to the *in vivo* microenvironment for cells to grow into various tissues. Silk 3D porous scaffolds have been prepared using freeze drying, porogen leaching and solid freeform fabrication techniques. 3D porous matrices with good control of pore size, porosity and structural content can be formed from reprocessed silk fibroin, either via organic solvent or aqueous processes, and using either salt leaching, gas foaming or freeze-drying techniques (Nazarov *et al.*, 2004; Kim *et al.*, 2005). The nature of the process used to form the porous matrices directly influences mechanical properties, degradability, and cell and tissue formation when used as scaffolds in tissue engineering. Pore sizes in these scaffolds

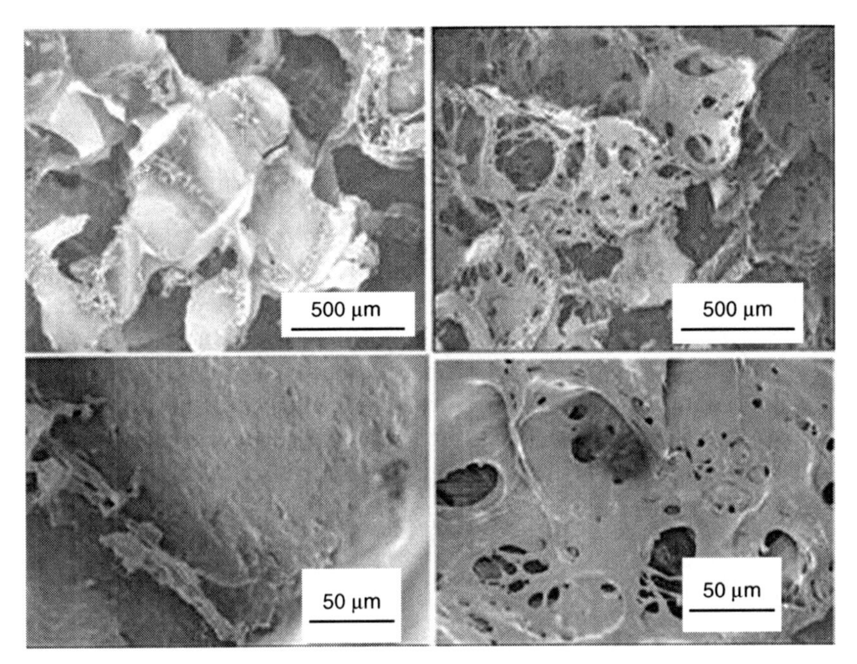

6.3 Scanning electron microscopy images of 3D porous scaffolds formed from reprocessed fibroin of *B. mori.* Top images show scaffolds produced using organic solvent processing (HFIP); bottom images show scaffolds produced using all-aqueous processing. Left and right images are taken at different locations.

can be controlled from below 100 mm diameter to above 1000 mm, depending on processing conditions (Fig. 6.3).

Aqueous-based porous silk sponges can be prepared using variable size salt crystals as porogen, with control of pore sizes from 490 to 940 mm, by manipulating the percentage of silk solution and size of salt crystals. Pore sizes are 80–90% smaller than the size of salt crystals due to the limited solubilization of the surface of the crystals during super-saturation of the silk solution prior to solidification. Aqueous silk fibroin sponges demonstrate improved cell attachment than the solvent-based porous sponges, probably due to these rougher surfaces.

Porous three-dimensional silk sponges have been utilized in a number of studies with cells to generate various connective tissues. Tissue engineered silk sponges were found useful for healing critical size femur defects in rats. Aqueous porous sponges with large pore sizes (900 mm) can be used for bone tissue engineering. Aqueous-based silk fibroin sponges seeded with chondrocytes also support cartilage tissue engineering. Due to the success in achieving good control over porosity and pore size, porogen leached 3D silk scaffolds have been used for bone and cartilage tissue engineering applications.

6.2.5 Membranes

Membranes (e.g. using reverse osmosis) are used in processes such as desalination of water, water purification, the bioprocessing industry and some chemical processes. Pure sericin cannot easily be made into membranes, but membranes of sericin cross-linked, blended or copolymerized with other substance are made more easily, because sericin contains large amount of amino acid with neutral polar functional groups. Sericin containing membranes are quite hydrophilic. Silk fibroin membranes have been used to separate water and alcohol (Chisti, 1998). Mizoguchi *et al.* (1991) describe a cross-linked thin film made of sericin for use as a separating membrane for water and ethanol. Acrylonitrile used in making certain synthetic polymers can be copolymerized with sericin to prepare a protein containing synthetic polymer film for separating water from organics (Yamada *et al.*, 1993; Yamada and Fuwa, 1994).

6.3 Silk non-wovens

Silk non-woven fabrics can be developed from silk reeling waste and hard waste generated during twisting and weaving on shuttle-less looms. In the process of conversion of cocoon to fabric, about 4000 MT of silk waste of different forms is being generated, annually (in India alone). At present, this waste is used for manufacture of spun silk yarn, noil yarn, throwster yarn and carpet yarn, besides hand-spun yarn. This waste can be more effectively utilized for development of silk non-woven fabrics for diversified end-uses. In addition, the hard waste generated to the extent of 300 MT in silk twisting and weaving by 100% export oriented units (EOUs) is not being used for any value added purpose except in manufacture of coarser yarn for carpets. The same may effectively be utilized for development of non-woven silk fabrics. The other wastes like pierced and cut cocoons may also be tried. The web formation by air laid method and bonding by chemical/needle punching may be attempted for production of no-woven webs. Based on the end-uses, non-woven fabric of specific weight can be produced for various applications including technical and medical textiles. Non-woven silk fabrics have applications in such areas as inner lining for warm garments, garments and blankets, carpets and wall covering, automotive carpeting and insulation.

6.4 Fluorescent silks

The world's first silks exhibiting fluorescence and other pioneering properties have been successfully developed as a result of transgenic silkworm research conducted by Japanese researchers. Researchers have developed

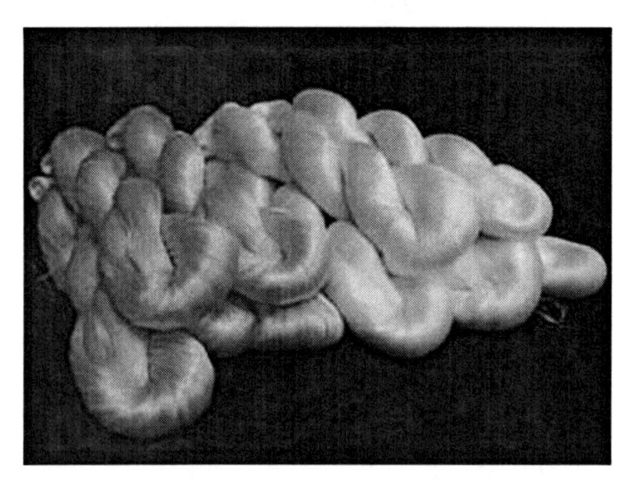

6.4 Fluorescent silk yarns.

three lines of transgenic silkworms. The first line produces silk threads that emit green, red or orange fluorescent light. These threads were created by introducing genes into silkworm eggs that promote the generation of fluorescent proteins. It has been possible to achieve green fluorescence using genes extracted from jellyfish, a technique developed by Nobel Prize-winning chemist Osamu Shimomura, and red and orange fluorescence with genes extracted from coral, a technique that has already been used in commercial applications. The fluorescent silk threads (Fig. 6.4) have great potential for use in the fashion industry, and there is expected to be considerable demand for them from producers of high-end apparel.

In another invention at the Institute of Materials Research and Engineering (IMRE), Singapore, silkworms that produce vibrantly coloured and luminescent silks have been created. The resulting fibre offers a cheap way to circumvent the dyeing process and may even have medical applications (Dumas and Pickrell, 2011). By feeding silkworms a mulberry mixture containing fluorescent dye, scientists were able to harvest brightly coloured and luminescent silk. The dye molecules are ingrained within the silk filaments to create permanent colour. This is done by adding the dye into the silkworm diet for the last 4 days of the larva stage. The resulting coloured cocoon (Fig. 6.5) can then be harvested and processed using normal processes. The process provides a more sustainable method of dyeing silk for the silk industry by reducing the vast amounts of water and dyes used in the labour-intensive conventional dyeing process. The technology is simple and cheap enough to be translated to an industrial scale and the researchers are currently working with potential industry partners to scale up the process and bring a product to market within a few years. The research team also envisages creating

6.5 Feeding silkworms with mulberry containing a fluorescent dye to produce a coloured cocoon.

silk with anti-bacterial, anti-coagulent and anti-inflammatory properties that could be used in wound dressing or even as biomedical frameworks for repairing damaged tissues. Silk wound dressings could also be created that have compounds with monitoring or sensing capabilities.

6.5 Biomedical applications of silk

Silkworm silk fibres have long been used in biomedical applications, particularly as sutures, given their high tensile strength, controllable bio-degradability, haemostatic properties, non-cyto toxicity, low anti-genicity and non-inflammatory characteristics (Li *et al.*, 2003; Jin *et al.*, 2004; Mauney *et al.*, 2007). The ability to control molecular structure and morphology through processing and surface modification options have expanded its use in a wide range of medical applications (Hakimi *et al.*, 2007).

6.5.1 Sutures and wound healing

Silk from the silkworm, *B. mori*, has been used as biomedical suture material for centuries. The unique mechanical properties of these fibres provide important clinical repair options. Silk has been widely used in sutures for wound ligation, surpassing collagen over the past 100 years (Altman *et al.*, 2003). Silk sutures are also used in ocular, neural and cardiovascular surgery. Silk's knot strength, handling characteristics, and the ability to rest close to the tissue surface make it a popular suture in cardiovascular applications. Silk sutures have a relatively very high modulus of elasticity. The moisture regain of a typical silk fibre is about 9.9% (Robson, 1985). Silk sutures also have very good handling properties with excellent knot security.

However, silk sutures are known to have undesirable *in vivo* loss of tensile strength and somewhat higher tissue reactions and in-growth. There have been a number of attempts to remedy these problems. Silk sutures have been impregnated with a hydrophobic thermoplastic elastomer to improve their performance (Shalaby *et al.*, 1984). Impregnated silk sutures provide better tensile breaking strength retention and significantly reduced tissue reactions and cellular invasion. Altman *et al.* (2003) have suggested that the higher tissue reactions to silk fibres may be attributed to residual sericin, although silk fibroin fibres have comparable biocompatibility to other natural materials such as collagen. Various types of surface treatments are used to render silk non-capillary, serum proof, or resistant to the in-growth of tissues. Wax or silicone has been used as the coating material. Tissue in-growth is prevented by encasing the twisted silk fibres in a non-absorbing coating of tanned gelatin or other protein substances.

Silk has also been used in wound healing. Wu *et al.* (1996) studied the properties and application of wound protective membrane made by silk fibroin, concluding that the fibroin membrane has good wound healing properties. Tasubouchi (1999a) developed a silk fibroin-based wound dressing that could accelerate healing and could be peeled off without damaging the newly formed skin. The non-crystalline fibroin film of the wound dressing had a water content of 3–16% and a thickness of 10–100 μm. Subsequently, the wound dressing was made with a mixture of both fibroin and sericin (Tsubouchi, 1999b). A membrane composed of sericin and fibroin is an effective substrate for the proliferation of adherent animal cells and can be used as a substitute for collagen. Minoura *et al.* (1995) and Tsukada *et al.* (1999) investigated the attachment and growth of animal cells on films made of sericin and fibroin. Cell attachment and growth were dependent on maintaining a minimum of around 90% sericin in the composite membrane. Film made of sericin and fibroin has excellent oxygen permeability and is similar to human cornea in its functional properties. It hoped that the sericin–fibroin blended film could be used to form article corneas (Murase, 1994). A novel mucoadhesive polymer has been prepared by template polymerization of acrylic acid in the presence of silk sericin (Ahn *et al.*, 2001). Silk protein can be made into a biomaterial with anticoagulant properties, by a sulfonation treatment of sericin and fibroin (Tamada, 1997).

6.5.2 Tissue engineering

Silk fibroin in various formats (films, fibres, nets, meshes, membranes, yarns and sponges) has been shown to support stem-cell adhesion, proliferation and differentiation *in vitro* and promote tissue repair *in vivo*. In particular, stem-cell-based tissue engineering using 3D silk fibroin scaffolds has expanded the use of silk-based biomaterials as promising scaffolds for engineering a range of skeletal tissues like bone, ligament and cartilage, as well as connective tissues like skin (Wang *et al.*, 2006). Recent studies suggest that silk fibroin fibres exhibit comparable biocompatibility *in vitro* and *in vivo* with other commonly used biomaterials such as polylactic acid and collagen (Altman *et al.*, 2003). Fibroin hydrogels have been proposed as scaffolds able to promote *in situ* bone regeneration (Matta *et al.*, 2004).

Hu (2006) reported that the recombinant human-like collagen (RHLC) can be blended with fibroin to prepare a novel biocompatible film as a scaffold material for hepatic tissue engineering applications. Solution blending is used to incorporate RHLC with silk fibroin to enhance the blend films biocompatibility and hydrophilicity while maintaining elasticity. Soluble fibroin enhances insulin sensitivity and glucose metabolism in 3T3-L1 Adiposities. The fibroin protein is one kind of biological materials used for artificial skin and others medical application. Silk fibroin membrane supports the application as photo sensor for hydrogen peroxide analysis. Silk protein sericin suppresses DMBA-TPA induced mouse skin tumour genesis by reducing oxidative stress, inflammatory responses and the endogenous tumour promoter TNF-alpha (Zhaorigetu *et al.*, 2003).

Nanofibrous silk fibroin chondrogenic scaffolds have been demonstrated to have excellent potential in cartilage tissue engineering. The surface modification of these scaffolds, such as microwave-induced argon plasma treatment, induced the enhanced attachment and proliferation of human articular chondrocytes, with a concomitant increase in the glycosaminoglycan synthesis (Baek *et al.*, 2008). Silk fibroin was introduced into amino polysaccharide chitosan complex scaffolds to delay matrix degradation when in contact with lysozyme solution (Bhardwaj and Kundu, 2011). The resultant complex structure also possessed a higher compressive strength and modulus compared to respective parameters of the individual scaffolds, and was suitable for the attachment and propagation of feline fibroblasts. Regenerated silk fibroin matrices have also been coated with a chitosan membrane to promote formation of crystalline vaterite disks of calcium carbonate on their surface (Wu *et al.*, 2011).

Osteoinductive and biodegradable composite biomaterials for bone regeneration were also fabricated from silk fibroin incorporated with silica particles (Mieszawska *et al.*, 2010). Human mesenchymal stem cells were found to successfully adhere, proliferate, and undergo osteogenic differentiation on silk/silica films, with evidence of early bone formation. Introduction of silica nanoparticles enhanced gene expression of bone sialoprotein and collagen type I osteogenic markers. Leaching of small sized silica particles from silk fibroin matrix further promoted formation of apatite deposits and enhanced collagen content.

6.5.3 Drug delivery

Sericin and fibroin have been recently explored in the field of drug delivery systems. Using fibroin-controlled release tablets, gels and mesosphere have been prepared. The applicability of fibroin, a major silk protein, to controlled release-type dosage tablets has been investigated *in vitro* and *in vivo*. Sulfated silk fibroins have anti-HIV-1 activity *in vitro*, apparently due to interference with the adsorption of virus particles to CD4+ cells, and completely blocked virus binding to the cells at a concentration of 100 micro gm/mL (Gotoh *et al.*, 2000).

Heparin bearing native silk fibroin powder blended with biomedical polyurethane were fabricated into composite membranes intended for the controlled release of heparin (Yang *et al.*, 2011). In addition to exhibiting excellent hydrophilicity, water vapour permeability and water absorption, the heparin release rate could be controlled by altering silk fibroin to polyurethane ratio, heparin content and membrane thickness. The delivered heparin was characterized by excellent bioactivity, with the system exhibiting potential for enhancement of polyurethane biocompatibility and haemocompatibility (Liu *et al.*, 2011).

6.6 Silk fibre-reinforced composites

Silk fibres can be incorporated as reinforcements into thermoplastic matrices to create composite materials with a high strain to failure rate. Silk yarn is easily available as a waste product from textile processing, making silk fibre-reinforced composites environmentally friendly and cost effective. Incorporating silk fibres into biopolymer matrices such as epoxy resins produces a 'green' bio-composite. The silk fibres can contribute significantly to impact resistance by ensuring both sufficient strength and good deformability. Novel short silk fibre-reinforced poly(butylenes succinate) (PBS) bio-composites can be prepared with varying fibre contents by a compression moulding method. Mechanical and thermal properties have been studied in terms of tensile and flexural properties, thermal stability and expansion. The results show that chopped silk fibres play a significant role as reinforcement for improving mechanical properties. Surface modification of the silk fibres improves interfacial bonding between the fibre and the matrix. The tensile and flexural properties of PBS matrix resin improve significantly by increasing the short fibre content in the composites, showing a maximum value at a fibre loading of 50 wt% (Das, 2010).

6.7 Conclusions

Silk has been transformed in recent decades from an apparel and furnishing textile world to a fibre with a growing web of high-technology applications.

Fundamental discoveries into how silk fibres are made have provided the basis for a new generation of applications, from medical devices and drug delivery to electronics. The development of silk hydrogels, films, fibres and sponges is making possible advances in photonics and optics, electronics, adhesives and microfluidics, as well as engineering of bone and ligaments. Applications could include degradable and flexible electronic displays for sensors that are biologically and environmentally compatible and implantable optical systems for diagnosis and treatment. Techniques for reproducing natural silk protein in the lab continue to advance. Silks are also being cloned and expressed in a variety of hosts, including *E. coli* bacteria, fungi, plants and mammals, and through transgenic silkworms. One day, efficient transgenic plants could be used to crop silk in much the same way that cotton is harvested today.

6.8 References and further reading

Ahn, J.S., Choi, H.K., Lee, K.H., Nahm, J.H. and Cho, S. (2001), Novel mucoadhesive polymer prepared by template polymerization of acrylic acid in the presence of silk sericin, *J. Appl. Polym. Sci.*, **80**, 274–280.

Allmeling, C., Jokuszies, A., Reimers, K., Kall, S. and Vogt, P.M. (2006), Use of spider silk fibres as an innovative material in a biocompatible artificial nerve conduit, *J. Cell. Mol. Med.*, **10**, 770–777.

Altman, G.H., Horan, R.L., Lu, H.H., Moreau, J., Martin, I., Richmond, J.C. and Kaplan, D.L. (2002), Silk matrix for tissue engineered anterior cruciate ligaments, *Biomaterials*, **23**, 4131–4141.

Altman, G.H., Diaz, F., Jakuba, C., Calabro, T., Horan, R.L., Chen, J., Lu, H., Richmond, J. and Kaplan, D.L. (2003), Silk-based biomaterials, *Biomaterials*, **24**, 401–416.

Arcidiacono, S. Mello, C.M., Butler, M., Welsh, E., Soares, J.W., Allen, A., Ziegler, D., Laue, T. and Chase, S. (2002), Aqueous processing and fiber spinning of recombinant spider silks, *Macromolecules*, **35**, 1262–1266.

Asakura, T., Yoshimizu, H. and Kakizaki, M. (1990), An ESR study of spin-labeled silk fibroin membranes and spin-labeled glucose-oxidase immobilized in silk fibrion membranes, *Biotechnol. Bioeng.*, **35**, 511–517.

Aslani, M.A. and Eral, M. (1994), Investigation of uranium recovery from dilute aqueous solutions using silk fibroin, *Biol. Trace Elem. Res.*, **43**, 737–743.

Baek, H.S., Park, Y.H., Ki, C.S., Park, J.-C. and Rah, D.K. (2008), Enhanced chondrogenic responses of articular chondrocytes onto porous silk fibroin scaffolds treated with microwave-induced argon plasma, *Surf. Coat. Technol.*, **202**, 5794–5797.

Bhardwaj, N. and Kundu, S.C. (2011), Silk fibroin protein and chitosan polyelectrolyte complex porous scaffolds for tissue engineering applications, *Carbohydrate Polymers*, **85**, 325–333.

Cao, Y. and Wang, B. (2009), Biodegradation of silk biomaterials, *Int. J. Molec. Sci.*, **10**, 1514–1524.

Chisti, Y. (1998), Strategies in downstream processing. In *Bioseparation and Bioprocesssing: A Handbook*, ed. Subramanian, G., New York: Wiley-VCH, 3–30.

Das, A. (2010), Silk reinforced composites, B.Tech. Thesis, Department of Mechanical Engineering, National Institute of Technology, Rourkela.

Demura, M., Asakura, T. and Kuroo, T. (1989), Immobilization of biocatalysts with *Bombyx mori* silk fibroin by several kinds of physical treatment and application to glucose sensors, *Biosensors*, **4**, 361–372.

Dumas, D. and Pickrell, J. (2011), Silkworms made to spin coloured silk, *Australian Geographic*, **March**, 5–7.

Freddi, G. and Tsukada, M. (1996), *Polymeric Materials Encyclopedia*, p. 7734, CRC Press, Boca Raton, FL.

Gellynck, K., Verdonk, P., Almqvist, K.F., Van Nimmen, E. and Ghey, T. (2005), Chondrocyte growth in porous spider silk 3D scaffolds, *Eur. Cell. Mater.*, **10**(2) 45.

Gole, R.S. and Kumar, P. (2006), Spider's silk: investigation of spinning process, web material and its properties, UG Project, Biological Sciences and Bioengineering, IIT Kanpur.

Gotoh, K., Izumi, H., Kanamoto, T., Tamada, Y and Nakashima, H. (2000), Sulfated fibroin, a novel sulfated peptide derived from silk, inhibits human immunodeficiency virus replication *in vitro*, *Biosci. Biotechnol. Related Articles, Books Biochem.*, **64**, 1664–1670.

Gulrajani, M.L. (2005), Sericin: a bio-molecule of value. Souvenir, 20th Congress of the International Sericultural Commission, Bangalore, 15–18 December, 21–29.

Hakimi, O., Knight, D.P., Vollrath, F. and Vadgama, P. (2007), Spider and mulberry silkworm silks as compatible biomaterials, *Composites*: *Part B*, **38**, 324–337.

Hu, K. (2006), Biocompatible fibroin blended films with recombinant human-like collagen for hepatic tissue engineering, *J. Bioact. Compat. Polym.*, **21**, 23–37.

Hudson, S.M. (1998), The spinning of silklike protein into fibers. In *Fibrous Proteins*, ed. McGrath, K. and Kaplan, D.L., Birkhouser, Boston, MA.

Huemmerich, D., U. Slotta and Scheibel, T. (2006), Processing and modification of films made from recombinant spider silk proteins, *Appl. Phys. A.*, **82**, 219–222.

Inouye, K., Kurokawa, M., Nishikawa, S. and Tsukada, M. (1998), Use of *Bombyx mori* silk fibroin as a substratum for cultivation of animal cells, *J. Biochem. Biophys. Meth.*, **18**, 159–164.

Jin, H.-J., Fridrikh, S.V., Rutledge, G.C. and Kaplan, D.L. (2002), Electrospinning *Bombxy mori* silk with poly(ethylene oxide), *Biomacromolecules*, **3**, 1233–1239.

Jin, H-J., Park, J., Kim, U.-J., Valluzzi, R., Cebe, P. and Kaplan, D.L. (2004), Biomaterials films of *Bombyx mori* silk with poly(ethylene oxide), *Biomacromolecules*, **5**, 711–717.

Junghans, F., Morawietz, M., Conrad, U., Scheibel, T., Heilmann, A. and Spohn, U. (2006), Preparation and mechanical properties of layers made of recombinant spider silk proteins and silk from silk worm, *Appl. Phys. A.*, **82**, 253–260.

Kato, N., Sato, S., Yamanaka, A., Yamadam, H., Fuwam, N. and Nomura, M. (1998), Silk protein, sericin, inhibits lipid peroxidation and tyrosinase activity, *Biosci. Biotechnol. Biochem.*, **62**, 145–147.

Kim, U.J., Park, J., Kim, H.J., Wada, M. and Kaplan, D.L. (2005), Three-dimensional aqueous-derived biomaterial scaffolds from silk fibroin, *Biomaterials*, **26**, 2775–2785.

Kim, U.-J., Park, J., Li, C., Jin, H.-J., Valluzzi, R. and Kaplan, D. L. (2004), Structure and properties of silk hydrogels, *Biomacromolecules*, **5**, 786–792.

Kundu, J., Patra, C. and Kundu, S.C. (2008), Design, fabrication and characterization of silk fibroin–HPMC–PEG blended films as vehicle for trans-mucosal delivery, *Mate. Sci. Eng.*, **28**, 1376–1380.

Li, C.M., Jin, H.J., Botsaris, G. and Kaplan, D.L. (2005), Silk apatite composites from electrospun fibers, *J. Mater. Res.*, **20**, 3374–3384.

Li, M., Ogiso, M. and Minoura, N. (2003), Enzymatic degradation behavior of porous silk fibroin sheets, *Biomaterials*, **24**, 357–365.

Liu, X., Zhang, C., Xu, W., Liu, H. and Ouyang, C. (2011), Blend films of silk fibroin and water-insoluble polyurethane prepared from an ionic liquid, *Mater. Lett.*, **65**, 2489–2491.

Matsumoto, A., Kim, H.J. Tsai, I.Y., Wang, X., Cebe, P. and Kaplan, D.L. (2006), Silk. In *Hand Book of Fibre Chemistry,* Third Edition, ed. Lewin, M., CRC Press, New York.

Matta, A., Migliaresi, C., Faccioni, F., Torricelli, P., Fini, M. and Giardino, R. (2004), Fibroin hydrogels for biomedical applications, preparation, characterization and *in vitro* cell culture studies, *J. Biomater. Sci. Polym.*, **15**, 851–864.

Mauney, J.R., Nguyen, T., Gillen, K., Kirker, C., Gimble, J.M. and Kaplan, D.L. (2007), Engineering adipose-like tissue *in vitro* & *in vivo* utilizing human bone marrow and adipose-derived mesenchymal stem cells with silk fibroin 3D scaffolds, *Biomaterials*, **28**, 5280–5290.

Maynes, E., Mann, S. and Vollrath, F. (1998), Fabrication of magnetic spider silk and other silk-fiber composites using inorganic nanoparticles, *Adv. Mater.*, **10**, 801.

Mieszawska, A.J., Fourligas, N., Georgakoudi, I., Ouhib, N.M., Belton, D.J., Perry, C.C. and Kaplan, D.L. (2010), Osteoinductive silk–silica composite biomaterials for bone regeneration, *Biomaterials*, **31**, 8902–8910.

Minoura, N., Aiba, S., Gotoh, Y., Tsukada, M. and Imai, T. (1995), Attachment and growth of cultured fibroblast cells on silk protein matrices, *J. Biomed. Mat.*, **29**, 1215–1221.

Mizoguchi, K., Iwatsubo, T. and Aisaka, N. (1991), Separating membrane made of cross-linked thin film of sericin and production thereof, Japan Patent 03-284337A.

Muller, W.S., Samuelson, L.A., Fossey, S.A. and Kaplan, D.L. (1993), Formation and characterization of Langmuir silk films, *Langmuir*, **9**, 1857–1861.

Murase, M. (1994), Method for solubilizing and molding cocoon silk, artificial organ made of cocoon silk, and medical element made of cocoon silk, Japan Patent 06-166850A.

Nazarov, R., Jin, H.J. and Kaplan, D.L. (2004), Porous 3-D scaffolds from regenerated silk fibroin, *Biomacromolecules*, **5**, 718–726.

Padamwar, M.N., Pawar, A.P., Daithankar, A.V. and Mahadik, K.R. (2005), Silk sericin as a moisturizer an *in vivo* study, *J. Cosmet. Dermat.*, **4**, 250–257.

Rammensee, S., Huemmerich, D., Hermanson, K.D., Scheibel, T. and Bausch, A.R. (2006), Rheological characterization of hydrogels formed by recombinantly produced spider silk, *Appl. Phys. A.*, **82**, 261–264.

Robson, R.M. (1985), Silk: composition, structure, and properties. In *Fiber Chemistry, Handbook of Fiber Science and Technology: Vol IV*, ed. Lewin, M. and Pearce, E.M., New York, Marcel Dekker, Chapter 7.

Shalaby, S.W., Stephenson, M., Schaap, L. and Hartley, G.H. (1984), Composite sutures of silk and hydrophobic thermoplastic elastomers, US Patent 4,461,298.

Tamada, Y. (1997), Anticoagulant and its production, Japan Patent 09-227402A.

Tsubouchi, K. (1999a), Wound covering material, US patent 5951506.

Tsubouchi, K. (1999b), Occlusive dressing consisting essentially of silk fibroin and silk sericin and its production, Japan Patent 11-070160A.

Tsukada, M., Hayasaka, S., Inoue, K., Nishikawa, S. and Yamamoto, S. (1999), Cell culture bed substrate for proliferation of animal cell and its preparation, Japan Patent 11-243948A.

Wang, Y., Kim, H.-J., Vunjak-Novakovic, G. and Kaplan, D.L. (2006), Stem cell based tissue engineering with silk biomaterials, *Biomaterials*, **27**, 6064–6082.

Wu, C.Y., Tian, B.Z., Zhu, D., Yan, X.M., Chen, W. and Xu, G.Y. (1996), Properties and application of wound protective membrane made from fibroin, International Silk Congress, Suzou Institute of Silk Technology.

Wu, Y., Cheng, C., Yao, J., Chen, X. and Shao, Z. (2011), Crystallization of calcium carbonate on chitosan substrates in the presence of regenerated silk fibroin, *Langmuir*, **27**, 2804–2810.

Yamada, H., Fuwa, N. and Nomura, M. (1993), Synthetic fiber having improved hygroscopicity, Japan Patent 05-339878A.

Yang, H.-J., Xu, H.-Y., Zhu, G.-C., Ouyang, C.-X., Wang, X.-G. and Xu, W.-L. (2011), Composite membranes of native silk fibroin powder and biomedical polyurethane for controlled release of heparin. *Proceedings of the Institution of Mechanical Engineers, Part H: Journal of Engineering in Medicine*, **225**, 421–433.

Yoshimizu, H. and Asakura, T. (1990), Preparation and characterization of silk fibroin powder and its application to enzyme immobilization, *J. Appl. Polymer Sci.*, **40**, 127.

Yoshimura, T., Shimizu, Y., Kurotani, W., Yamaoka, R. and Hayashiya, K. (1989), Application of fibroin membrane to immobilizing coenzed insect cell culture for use as vaccine, *Agri. Biol. Chem.*, **52**, 3201–3202.

Zhang, F., Zhao, Y., Chen, X., Xu, A.Y., Huang, J.T. and Lu, C.D. (1999), Fluorescent transgenic silkworm, *Acta Biochem. Biophys.*, **23**, 119.

Zhaorigetu, S.N., Sasakim M., Watanbe, H. and Kato, N. (2003), Silk protein, sericin, suppresses DMBA-TPA induced mouse skin tumorigenesis by reducing oxidative stress, inflammatory responses and endogenous tumor promoter TNF-alpha, *Oncol. Rep.*, **10**, 537–543.

7
Spider silks and their applications

DOI: 10.1533/9781782421580.156

Abstract: This chapter reviews the types and composition of spider silk, including amino acid composition. It also discusses the molecular structure of spider silk and its tensile properties.

Key words: spider silk, dragline silk, amino acid composition, tensile properties.

7.1 Introduction

In nature a wide range of materials have useful properties. One such material that has been exploited for millennia is silk (Gerritsen, 2002; Blackledge *et al.*, 2009; Heim *et al.*, 2010; Sutherland *et al.*, 2010). Silk from the larvae of the moth *Bombyx mori* (silkworms) is processed on a large scale in silkworm farms for textile applications. Silk from spiders has also been utilized throughout history. Scientists and engineers have long envied spiders' ability to manufacture a material that is simultaneously strong, fine and tough. This combination of properties makes spider silk an extremely attractive fibre for numerous applications in medicine, defence and the leisure industry. However, in contrast to silk from silkworms, spider silk has not been exploited on an industrial scale (Bunning, 1994).

Spider silk is a natural filamentous protein fibre. Many spider silks have better mechanical properties than silkworm silk (Gosline, 1999). The different properties are due to the different natural functions of the material. Silkworms use their silk for protection during their metamorphosis, while many spiders use silk to catch prey. Of particular interest are the silks from spiders that produce orb webs which are used to catch aerial prey. The silk in these webs needs to be capable of capturing and holding the spider's flying prey, which requires interplay of various silks with different properties. Female orb-weaving spiders can produce up to seven different silks with a range of properties. Spider silks are characterized by remarkable diversity in their chemistry, structure and functions, ranging from orb-web construction to adhesives and cocoons. To date, most researchers have focused their

7.1 Dragline silk produced by spiders.

attention on 'dragline' silk, used by spiders as a safety line and also as the frame for their webs (Fig. 7.1; Gould, 2002).

There are a number of reasons why spider silk has not been cultivated on the same scale as silk from silkworms. This is due to:

- difficulty in raising dense populations of spiders due to their solitary and predatory nature;
- spider webs not being reelable as a single fibre, unlike the fibroin from the cocoon of the silkworm;
- spiders being able to generate only small quantities of silk (about 137 m of fibre can be reeled from the ampullate gland of a fibre and only 12 m of silk is found in a complete web) in comparison to the silkworm cocoon silk;
- the fact that several types of silk are utilized in web construction from a single spider.

The best option for developing spider silk use is developing an alternative biotechnological means of generating spider silk (Kluge *et al.*, 2008). Recent progress in genetic engineering of spider silks and the development of new chimeric spider silks with enhanced functions and specific characteristics have advanced spider silk technologies.

7.2 Types of spider silk

Spider webs can take a variety of forms but the most common type is the orb-web. Different families of spiders *Araneus, Nephila* builds orb webs and other families of spiders construct tangle and sheet webs. The golden orb web-spider is the most spectacular and most investigated of all spiders (Fig. 7.2). They measure around 5 cm in length without the legs and up to 25 cm with stretched legs. These spiders are known to make a large,

7.2 Golden orb spider (*Nephila clavipes*).

golden-coloured web. These webs can be 2 m in diameter and some tribes use these webs as fishing nets (Champion de Crespigny *et al.*, 2001).

The construction of the orb-web is a feat of engineering involving material tailoring, optimization of material interfaces and conservation of resources to promote survival of the spider (Denny, 1980; Gosline, 1986). In addition, the web absorbs water from the atmosphere, and ingestion by the spider may provide a significant contribution to water intake needs. Around 70% of the energy is dissipated through viscoelastic processes upon impact by a flying insect into the web (Gosline, 1986). Thus the web balances stiffness and strength against extensibility, both to keep the web from breaking and the insect from being ejected from the web by elastic recoil (Gosline, 1986). The ability to dissipate the kinetic energy of a flying insect impacting the web is based on the hysteresis of radial threads and also aerodynamic damping by the web (Edmonds and Vollrath, 1992; Lin *et al.*, 1995). Some orb webs appear to be at least in part recycled by ingestion as a conservation tool, and some of the amino acids are reused in new webs.

Properties of spider silk vary by species and by function. Depending upon the species, an individual spider has more than five different silk glands where silks of different properties are produced before extrusion through three pairs of spinnerets. The dragline originates in the ampullate gland and is extruded through the anterior spinnerets. Spider dragline is the strongest and toughest of the silks a spider makes, and is the structural framework of a spider web. Orb-weaving spiders are able to synthesize as many as seven different types of silks by specialized glands (Table 7.1). The information contained within the silk proteins for each of these different types of silks provides important insights into protein structure–function relationships.

Table 7.1 Different types of spider silks

Serial no.	Silk gland	Use	Spinneret
1	Major ampullate	Dragline, frame threads	Anterior
2	Minor ampullate	Dragline reinforcement	Median
3	Pyriform	Attachment disk	Anterior
4	Aciniform	Swathsing prey	Median, posterior
5	Cylindrical (tubuliform)	Cocoon construction	Median, posterior
6	Aggregate	Sticky silk glue for capture spiral	Posterior
7	Flagelliform	Capture spiral	Posterior

Spider silks are used for nearly all aspects of a spider's life including capturing and swathing prey or protecting the cocoon, to name just a few.

7.3 Chemical composition

The molecular structure of silk consists of regions of protein crystals separated by less organized protein chains. The primary structural modules give rise to diverse secondary structures that, in their turn, direct functions of different silks. As the most heavily studied secondary structure of silks, crystalline β-sheets contribute to the high tensile strength of silk fibres. β-sheets form through natural physical cross-linking of amino acid sequences, which in spider and silkworm silk consist of multiple repeats of mainly alanine, glycine–alanine or glycine–alanine–serine.

The non-crystalline regions of silk are commonly made up of (Kaplan *et al.*, 1994, 1998):

- β-spirals similar to a β-turn composed of GPGXX repeats (where X is mostly glutamine);
- helical structures composed of GGX.

These semi-amorphous regions provide silk with elasticity. For example, the flagelliform silk from Nephila clavipes is rich in the GPGXX motif, and this sequence results in a highly elastic fibre that functions in prey capture. In addition to the crystalline and semi-amorphous regions, non-repetitive regions are present at the amino- and carboxyl termini of the proteins. Although the impact of these termini on mechanical response is not fully understood, it has been speculated that they might play a role in the controlled assembly of silk proteins (Van Beek, 2002; Scheibel, 2004).

Each silk-producing spider synthesizes silk proteins that offer a rich diversity of primary sequences and secondary structures. For example, the common amino acid modules in the silk fibres synthesized by the Araneomorphae (true spiders) can be grouped into four categories: poly–Ala, poly–Ala–Gly,

GPGXX, GGX and a spacer sequence (Hayashi, 1999). Most recently, Garb and colleagues characterized six novel silk proteins from the Mygalomorphae (tarantulas) that do not contain these four categories found in true spider spidroins. These newly characterized tarantula silks, as a result, do not possess high tensile strength and elasticity. This finding supports the hypothesized role of poly–Ala and GPGXX modules in forming dragline and flageliform silks (Garb, 2007).

In most of the above cases, the fundamental process of silk protein self-assembly into functional materials remains consistent, with the more hydrophobic domains, mainly the alanine, glycine–alanine and glycine–alanine–serine repeats, driving the process. In most spider silks, β-sheet formation is achieved in a spinning duct caused by the progressive loss of water in the gland and alignment of the hydrophobic regions during flow (Fig. 7.1; Vollrath, 2001; Jin, 2003; Bini, 2004; Huemmerich, 2004; Exler, 2007). Exceptions are the more hydrophilic silks, such as those involved in adhesion, wherein charge interactions can play a more dominant role than the hydrophobic interactions. The self-assembly without chemical cross-linking provides stability while still allowing enzymatic digestion (Scheibel, 2004) or slow degradation under appropriate environmental conditions (Horan, 2005).

The dragline silk from the spider *N. clavipes* originates from the major ampullate gland in the abdomen and contains two proteins or spidroins called major ampullate silk protein 1 (MaSp1) or spidroin 1, with a molecular weight around 275 000 Da, and a second similar protein but enriched with proline (Xu and Lewis, 1990; Hinman, 1992). There are no sericin-like proteins associated with the dragline fibre. MaSp1 contains amino acid repeats considerably shorter than those found in the silkworm fibroin and not as highly conserved. The repeats consisting of polyalanines of 6–12 residues are responsible for the formation of the β-sheet crystals in an analogous fashion to the GA repeats for the silkworm. Regions in the spider dragline silk with GGX repeats – where X = alanine, tyrosine, leucine or glutamine – are involved more in the flexible regions of the proteins when formed into fibres. It is worth noting that the distinctions between silkworm and spider repeat sequences responsible for β-sheet crystal formation are becoming blurred as more species of these types of organisms become sequenced and somewhat similar repeats can be identified both in silkworms and in spiders.

7.4 Amino acid composition and molecular structure of dragline silk

The major and minor ampullate fibroins contain the highest levels of glycine and alanine relative to other spider silk family members. These levels approach >50% of the total amino acid content (Lombardi and Kaplan, 1990; Casem *et al.*, 1999). Biochemical experiments show dragline silk is a composite material largely composed of two structural proteins or spidroins (contraction of the words spider and fibroin) called MaSp1 and MaSp2 (Xu and Lewis, 1990; Hinman and Lewis, 1991). Structural studies demonstrate the major ampullate spidroins form the core of the fibre that is wrapped inside a glycoprotein coat. Although the identities of the constituents of the glycoprotein layer remain unknown, experimental evidence supports the belief that this layer is added in the ampulla prior to extrusion (Casem *et al.*, 2002; Sponner *et al.*, 2005).

The molecular sequences coding for the dragline silk fibroins were the first to be identified from *N. clavipes* (Xu and Lewis, 1990). Recently, the complete genetic blueprints for MaSp1 and MaSp2 were determined from *L. hesperus* (Ayoub *et al.*, 2007). The predicted sequences for these fibroins encode large molecular weight proteins that are approximately 3500 amino acids in length. These spidroins are highly modular, each containing internal repetitive block repeats that are flanked by N- and C-terminal non-repetitive ends comprising approximately 100 amino acids. The internal block repeats are rich in glycine and alanine; these regions form polyalanine or polyalanine–glycine stretches that are interrupted by glycine–rich regions. The polyalanine segments form β-sheet crystal domains and are responsible for the high tensile strength while the glycine-rich regions adopt 31-helix type structures and beta-turns that link the crystalline domains (Simmons *et al.*, 1996). These interconnecting glycine-rich regions constitute the semi-amorphous regions and have been implicated in the extensibility of the fibres. Extensibility of dragline silk fibres also has been attributed to glycine–proline–glycine–X–X (GPGXX) repeats within the MaSp2 protein sequence and the formation of beta-spirals. MaSp2 proteins have been shown to be tightly packed in certain core regions of fibres from *N. clavipes*, whereas MaSp1 appears to be uniformly distributed along the radial axis (Sponner *et al.*, 2005). These data demonstrate that MaSp1 and MaSp2 are not evenly distributed down the long axis of natural fibres. The biochemical mechanisms that modulate their differential localization are not well understood, but could be explained by differences in expression levels and/or their protein sequences that control partitioning during extrusion.

Interestingly, in natural fibres, MaSp1 and MaSp2 ratios have been shown to vary between different species (Tso *et al.*, 2005). These differences have been linked to the diet and environment of spiders (Craig *et al.*, 2000). It has been suggested that the synthesis of MaSp2 is more energetically expensive because of the higher cost associated with proline biosynthesis. Therefore, it would appear that spiders synthesize cheap silk when resources are limited, perhaps producing fibres that contain predominantly MaSp1 molecules. Testing the mechanical properties of composite spider silk fibres spun from regenerated dragline silk fibroins combined with different ratios of recombinant MaSp1 and MaSp2 will help clarify the structural roles of the MaSp1 and MaSp2 fibroins. It will also reveal how this spider silk composite

Table 7.2 Amino acid composition (mole%) of spider dragline silk and other protein fibres

Amino acid	Silkworm silk	Wool fibre	Spider silk
Glycine	43.7	8.4	37.1
Alanine	28.8	5.5	21.1
Valine	2.2	5.6	1.8
Leucine	0.5	7.8	3.8
Isoleucine	0.7	3.3	0.9
Serine	11.9	11.6	4.5
Threonine	0.9	6.9	1.7
Aspartic acid	1.3	5.9	2.5
Glutamic acid	1.0	11.3	9.2
Phenylalanine	0.6	2.8	0.7
Tyrosine	5.1	3.5	–
Lysine	0.3	2.6	0.5
Histidine	0.2	0.9	0.5
Arginine	0.5	6.4	7.6
Proline	0.5	6.8	4.3
Tryptophan	0.3	0.5	2.9
Cystine	0.2	9.8	0.3
Metheonine	0.1	0.4	0.4

behaves from a mechanical perspective. To a large extent, this has not been fully explored because obtaining large amounts of purified recombinant silk proteins for the spinning process is a difficult task.

Table 7.2 shows the amino acid composition of dragline silk in comparison with other protein fibres such as silkworm silk and wool keratin (Kapaln *et al.*, 1997). It may be observed that, in addition to fibroin, other components like glycoprotein, inorganic salts, sulphur containing amino acids and ionic forms of amines may be found in spider silk (Schulz and Toft, 1993; www/zoology.ubc.ca, 2005). Presence of these components play crucial roles like identification of species, regulation of water content of web, protection against micro-organisms. Presence of 12-methyltetradecanic acid and 14-methylhexadecanoic acid to the minor amounts impart antimicrobial properties to the spider silk. Wax-like esters are also present on the surface of spider silk.

In most spider silks, β-sheet formation is achieved in a spinning duct caused by the progressive loss of water in the gland and alignment of the hydrophobic regions during flow (Fig. 7.3) (Vollrath and Knight, 2001; Jin and Kaplan, 2003; Bini *et al.*, 2004; Huemmerich *et al.*, 2004; Exler *et al.*, 2007). Exceptions are the more hydrophilic silks, such as those involved in adhesion, wherein charge interactions can play a more dominant role than the hydrophobic interactions. The self-assembly without chemical cross-linking provides stability while still allowing enzymatic digestion (Scheibel, 2004) or slow degradation under appropriate environmental conditions (Horan *et al.*, 2005).

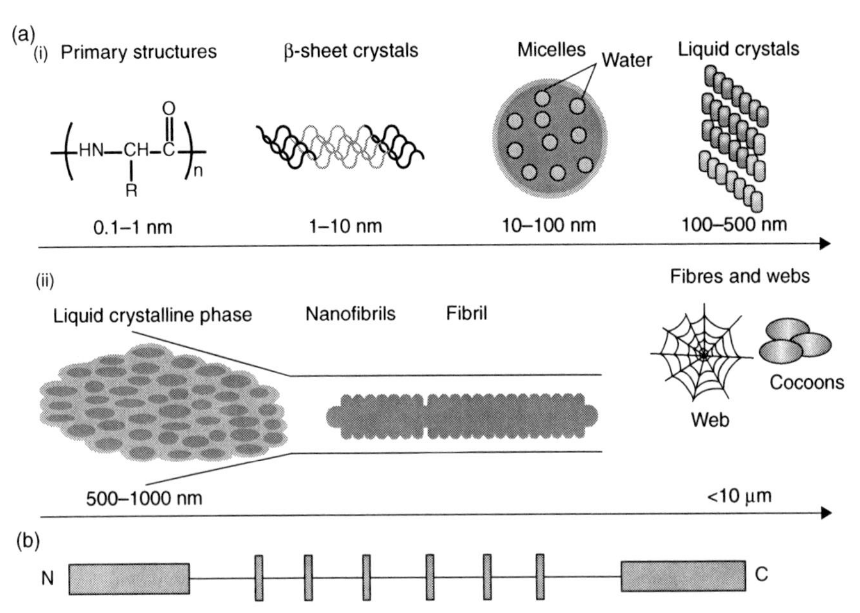

7.3 Structural hierarchy in the assembly of spider silk. (a) (i) Spider-silk proteins consist of repeats of amino acid sequences that self-assemble into β-sheets. This self assembly is driven by hydrogen bonding and also by hydrophobic regions. These interactions result in the formation of inter- or intra-molecular protein chain interactions. The β-sheet structures further assemble into soft micelles in a manner that excludes the hydrophilic ends to the perimeter. The interiors of the micelles contain water due to the presence of small 'spaces' that are more hydrophilic than the dominant hydrophobic domains. This does not represent a multi-shell structure; rather it is caused by the partititoning of the hydrophilic chain ends to the surface of the micelles and the location to the interior of the large and dominant hydrophobic domains and small hydrophilic spacers. With increasing protein concentration, micelles transform into gel-like states leading to metastable liquid crystalline structures. (ii) Triggers, such as physical shearing, or environmental factors, such as low pH, methanol, ultrasonication and electric fields, convert the gel states and liquid crystals into a more stable β-sheet structure. The resulting fibrils emerging from spinning ducts are combined into higher ordered structures as naturally constructed webs or cocoons. (b) Molecular structure of a spider-silk protein. In silks that are used for web architecture and other strong fibres, the underlying molecular structure consists mainly of hydrophobic regions (illustrated by thin black lines). These large hydrophobic regions are interspersed with small hydrophilic regions (illustrated by vertical rectangles) and are flanked by nonrepetitive domains at the N- and C-termini (horizontal rectangles).

7.5 General properties of spider silk

Spider silk is a semi-crystalline biopolymer with a unique combination of high tensile strength, high elasticity and high modulus. The 0.2–10 mm diameter silk fibres have a higher breaking energy than other natural or synthetic fibrous polymers, far exceeding that of high tensile steel and Kevlar on a weight-for-weight basis (Gosline *et al.*, 1986). Spider silk's unique combination of strength and elasticity is judged to be superior to that of synthetic high-tech fibres made of polyamide or polyester (Mukhopadhyay and Sakthivel, 2005). As well as being five times stronger than steel on a weight-by-weight basis, spider silk is finer than human hair, more resilient than any synthetic fibre, and completely biodegradable. Spider silk is shown to possess strength as high as 1.75 GPa at a breaking elongation of over 26%. It is three times tougher than aramid and other industrial fibres. Compared to silkworm silk, it is more waterproof and can absorb three times the impact force without breaking. All these properties are achieved in a fibre produced in ambient temperatures, low pressure and with water as a solvent.

In order to survive, the spider must use a minimum amount of silk in its web to capture prey, and yet the web must be able to stop and capture an insect flying at high velocity. To do this the web must absorb the energy of the insect without creating rebound which would trampoline the prey off the web. A recent study has concluded that the web and the spider silk used to construct it are nearly optimally designed for each other. The crucial factors in this optimization are tensile strength and elasticity. Spider silk thus combines strength and elasticity. This allows it to absorb more energy prior to breaking than any commonly used material (Lewis, 1992). Most synthetic fibres have either a high strength or a high elasticity, often higher than spider silk, but it is the combination of both that makes spider silk so special. Tensile tests of spider dragline, spider egg sac fibres and silkworm fibres (Fig. 7.4) show the differences in stress–strain behaviour of the different protein fibres. Spider egg sac silk has different stress–strain behaviour than that of spider draglines and silkworm cocoon fibres (Fig. 7.5).

Another fascinating evolutionary adaptation of dragline silk is its ability to super-contract. The fibre will contract to less than 60% of its original length when it is wetted. This results in nearly a 1000-fold decrease in the elastic modulus and an increased extensibility (Gosline and Denny, 1984). The practical application for the spider is that the web will tighten each day when it is wetted with dew, thus maintaining its shape and tension. There are several polymers which exhibit super-contraction in organic solvents but virtually none which will super-contract in water alone (Work, 1977b). This super-contraction is reversible and repeatable and can do mechanical work such as lifting a weight. An important feature in conjunction with super-contraction is the insolubility of the silks. Extreme chaotropic agents are required to solubilize any of the silks.

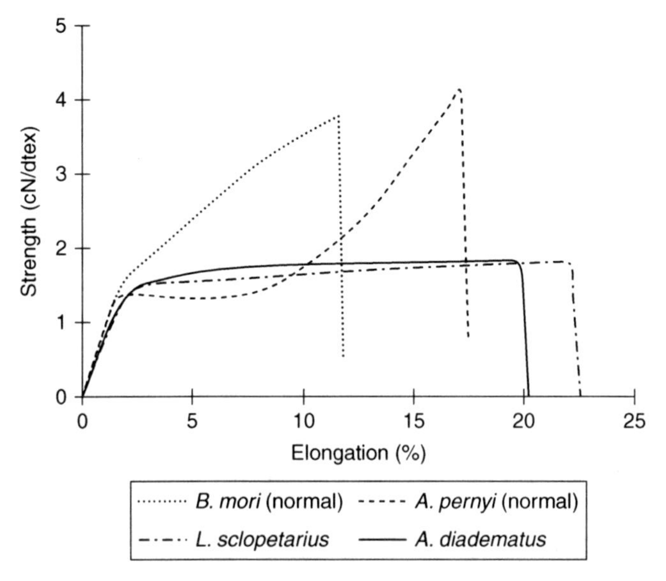

7.4 Stress–strain curves of cocoon silk: *B. mori* and *Antheraea pernyi* cocoon fibres; *Larinoides sclopetarius* and *Araneus diadematus* egg sac fibres.

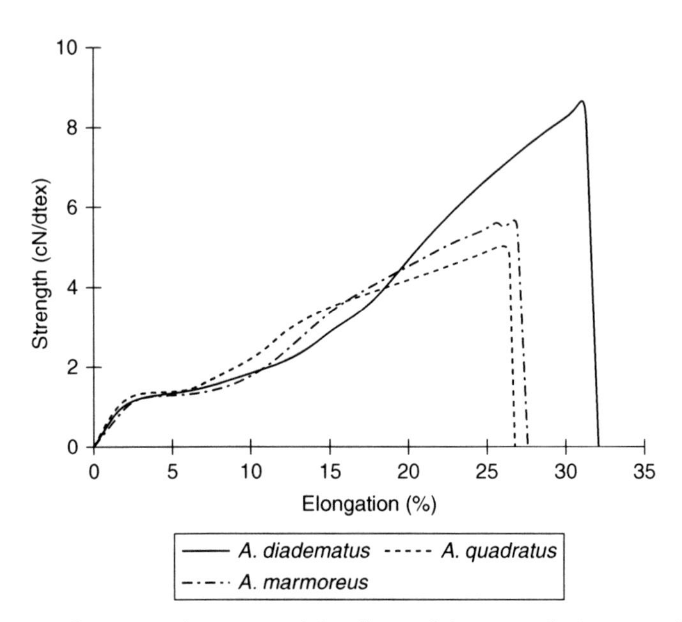

7.5 Stress–strain curves of draglines of *Araneus diadematus, Araneus quadrates* and *Araneus marmoreus.*

7.6 Tensile properties of spider silk

The tensile properties of spider silk vary from specimen to specimen, as demonstrated in the past studies of the Nephila clavipes spider by Ko *et al.* (2005). The spider silk was tested by simple elongation at a strain rate of 100% per minute using a gauge length of 1.25 cm. The stress–strain curve of the spider silk showed a sigmoidal shape (Fig. 7.6) indicating a balanced strength and elongation of 1.75 GPa (15.8 g/den) and 36%, respectively. A comparison of tensile properties of spider silk with other fibres is presented in Table 7.3. The stress–strain curve is characterized by three distinct regions: The 0–5% strain region characterized by a high initial modulus of 34 GPa, 5–21% strain region having a pseudo yield point at 5% before strain hardening to a maximum modulus of 22 GPa at 22% elongation and the 21–36% region exhibiting a gradual reduction of modulus until reaching a failure strength of 1.75 GPa. The area under the stress–strain curves showed a toughness level of 2.8 g/den. This is much higher than the toughness of the aramid fibre (0.26 g/den) and Nylon 6 fibre (0.9 g/den).

The stress–strain curve of the spider silk assumes a sigmoidal shape (Fig. 7.7) similar to that of an elastomer, demonstrating a good balance of strength and elongation at 1.75 GPa (15.8 g/den) and 36%, respectively. This 'rubber-like' stress–strain curve is characterized by three distinct regions: Region I (0–5%) is characterized by a high initial modulus of 34 GPa;

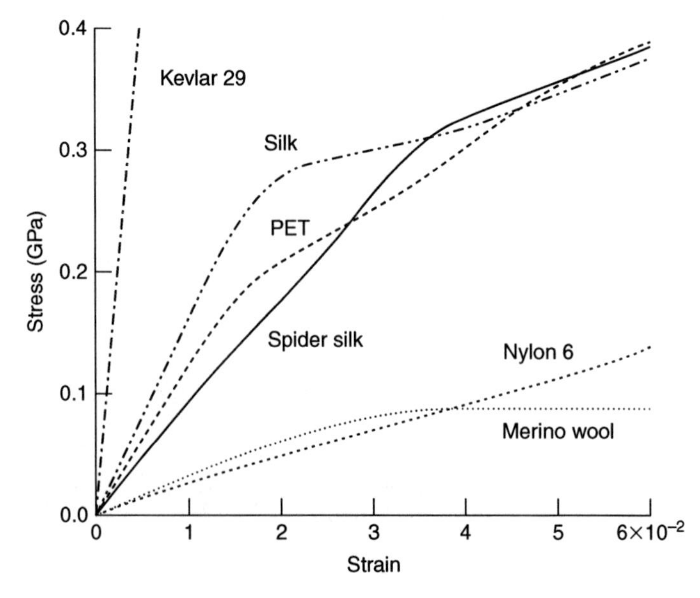

7.6 Tensile properties of spider silk compared with other fibres.

Table 7.3 Comparison of tensile properties of silk with other high-performance fibres

	Density (g/cm³)	Tenacity (GPa)	Extensibility (%)	Toughness (MJ/m³)
Nylon 6.6	1.1	0.95	18	80
Kevlar 49	1.4	3.6	3	50
Dragline of *A. diadematus*	1.3	1.1	27	160
Egg sac of *A. diadematus*	1.3	0.3	25–50	70
B. mori silk (mulberry silk)	1.3	0.6	18	70
Wool	1.3	0.2	50	60
Poly lactic acid (PLA)	1.24	0.7	22	90
Carbon fibre	1.8	4	1.3	25
High-tensile steel	7.8	1.5	1	6

Region II (5–21%) shows a pseudo yield point at 5% before strain hardening to a maximum modulus of 22 GPa at 22% elongation; and Region III (21–36%) exhibits a gradual reduction of modulus until reaching failure strength of 1.75 GPa at 36% elongation. An examination of the area under the stress–strain curves shows a toughness level of 2.8 g/denier. This is much higher than the toughness of the aramid fibre (0.26 g/denier) and Nylon 6 fibre (0.9 g/denier).

Kohler and Vollrath (1995) tested spider silk using a custom built, highly sensitive, rapid-response stress–strain gauge because of the small diameter and weak absolute force of single silk threads. The gauge was set up in such a way that fibres could be measured either in air or submerged full-length in a bath. The highly sensitive detection system with a FORT 10 force transducer (of World Precision Instruments) and linear extension mechanism (Pen Motor Assembly of Hewlett Packard) had a time resolution of a few ms and a force resolution of 30 N. The fibres were stretched in air as well as submerged in selected solvents and the nominal stress–strain characteristics of fibre calculated after normalization for fibre diameter and initial length. A comparative study done by Shao and Vollrath (1999) illustrates that four varieties of spider silk – Araneus, Latrodectus, Nephila and Euprosthenops – behave in quite a different fashion in terms of modulus, breaking strength, breaking elongation and strength. Euprosthenops shows the best values in terms of modulus, strength and shrinkage.

The stress–strain characteristics of dragline silks show large inter- and intra-specific differences between spiders in different families (Madsen *et al.*, 1999). Moreover, there can be a large daily variability in silk from individuals of the same species; and the spider's condition can (and frequently does) affect silk properties, starvation for example can lead to decreased breaking elongation. Speed of silk production also affects silk properties such

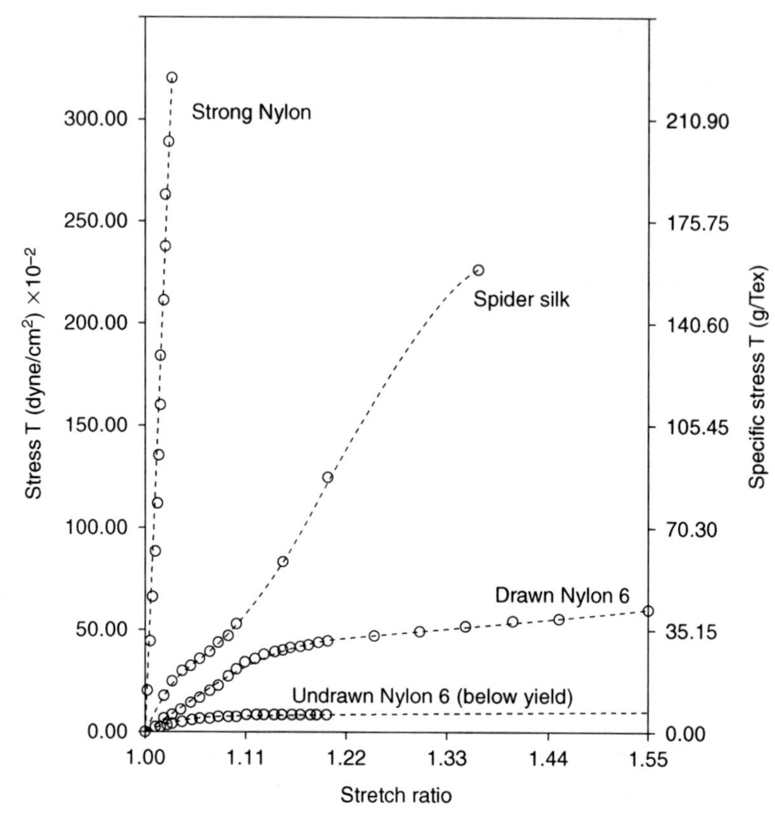

7.7 Tensile stress–strain curves of spider silk and other polyamide fibres.

that with increasing speed: (i) breaking elongation decreases; (ii) breaking stress increases; and (iii) Young's modulus increases. Finally, the spider's body temperature during silk production also plays an important role. Since spiders are ectotherms, this means that an individual can to some extent modify silk parameters by adjusting the time of building and the speed of running. However, the spider also seems to be able to modify silk parameters by direct nervous control (Madsen and Vollrath, 2000). As well as indirectly by its diet.

Based on microscopic evaluations of knotted single fibres, no evidence of kink-band failure on the compressive side of a knot curve form single dragline silk fibres was observed (Cunniff *et al.*, 1994). Synthetic high performance fibres fail by this mode even at relatively low stress levels; this is a major limitation with synthetic fibres in some structural composite applications (Kaplan *et al.*, 1998). It has been pointed out by Garrido *et al.* (2002) that mechanical characteristics of drag line spun during a vertical climb differ from the drag line spun when the spider crawls on a horizontal surface.

Also, the intrinsic stress–strain response of drag line spun during a vertical climb is significantly more reproducible than when this fibre is produced under other conditions.

Studies by Fritz Vollrath (1999) show that spider silk can be highly variable in its chemical composition and mechanical properties. Both external and internal conditions affect silk production and thus the mechanical properties of the finished thread. The protein sequences of major ampullate, minor ampullate and flagelliform silks from *N. clavipes* were characterized by Hayashi *et al.* (1999) and each structural element, termed a module, was associated with its impact on the mechanical properties of a silk fibre. In particular, correlations were drawn between an alanine-rich 'crystalline module' and tensile strength and between a proline-containing 'elasticity module' and extensibility.

The duct's convergent, or hyperbolic, geometry forces the dope flowing along it to elongate at a practically constant rate (Knight and Vollrath, 1999). As a result, the spherical droplets in the dope extend to form the long, thin and axially orientated canaliculi, which are thought to contribute to the thread's toughness (Shao *et al.*, 1999). The slow, constant nature of the elongation also ensures that only low and uniform stresses are generated. This prevents localized coagulation centres from forming prematurely (Knight and Vollrath, 1999) before the silk protein molecules in the dope have reached their optimal orientation. As in any other spinning, good molecular alignment contributes significantly to the thread's toughness (Northolt and Sikkema, 1991; Donald and Windle, 1992; Tirrell *et al.*, 1994; O'Brien *et al.*, 1998).

A much higher stress is generated during the rapid extension when the forming thread suddenly stretches, narrows and pulls away from the walls of the third limb of the duct. These high forces bring the dope molecules into alignment so that they are able to join together with hydrogen bonds to give anti-parallel beta conformation of the final thread. The spider's simultaneous and internal drawdown and material processing using phase-separation differs from industrial spinning, where the solvent escapes to the surface at the die's external opening. In the case of spider silk, the inside drawing process offers the obvious advantage: most of the water from the dope can be recycled by absorption from the duct. More importantly – the duct acts as a combined internal die and treatment bath (Vollrath *et al.*, 1998). The absolute size of the molecules and their size distribution are also important parameters affecting the toughness of the final thread – the spider silk containing predominantly a single large protein, with little variability of molecular weight (O'Brien *et al.*, 1998).

The extreme elasticity of this natural fibre comes from long spirals in the protein's configuration (www.sciencenews.org/sn_arc98/2_21_98/fob2.htm), propose researchers from the University of Wyoming in Laramie. The

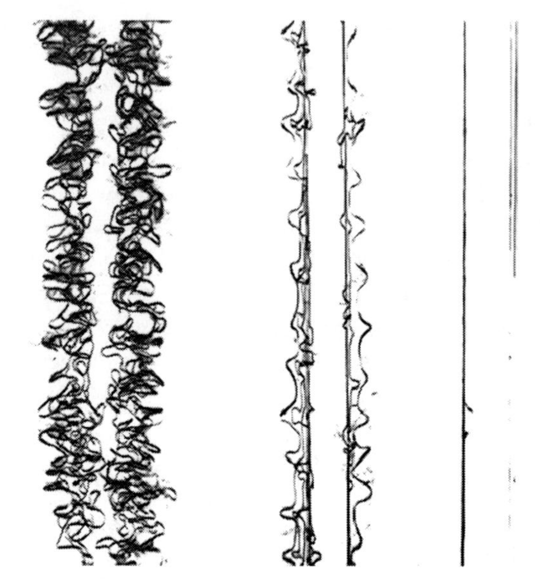

7.8 Strands of silk from a species of social spider, *Stegodyphus sarasinorum*, shown (from left to right): normal size, stretched to five times its original length and stretched to 20 times its original length.

helices present in the protein molecules act as molecular springs and make it elastic. A strand of spider silk (Fig. 7.8), of normal size, stretched five times and 20 times its original length showing its extensibility.

It has been found that capture silk protein from *N. clavipes* has a chain of thousands of amino acids having regions in which a sequence of five amino acids is repeated over and over, as many as 63 times. The researchers suggest that the segments of the protein with the repeating blocks form long, spring-like shapes. At the end of each five-amino-acid block, the protein kinks back on itself in a 180° turn. The series of turns eventually forms a spiral that 'looks exactly like a molecular spring'. Dragline silk proteins and capture silk proteins have similar turn-forming blocks of amino acids. However, these blocks repeat an average of 43 times in the capture silk, compared to only nine times in the dragline silk. That fivefold difference in length corresponds to the difference in elasticity between the two proteins.

7.7 Applications

Humans have been making use of spider silk for thousands of years. The ancient Greeks used cobwebs to stop wounds from bleeding and the Aborigines used silk as fishing lines and the people of the Solomon Islands still use silk as fish nets. Until the Second World War silk was used as the crosshairs in optical targeting devices such as guns and telescopes. There have been a number of more recent developments. Magnetic silk–fibre composites, for example, can be made by binding colloidal magnetite (Fe_3O_4) nanoparticles to threads of dragline spider silk (Maynes *et al.*, 1998). Such mineralized fibres retain their high strength and elasticity but can be oriented by an external magnetic field. Elemental mapping of sectioned fibres showed a dense, coherent surface coating of the iron oxide, which is partially retained after washing. Special silk–fibre composites can be obtained by similar procedures using conductive (metallic Au) or semiconductive (CdS) nanocolloidal precursors (Arcidiacono *et al.*, 2002). This study shows that organic/inorganic silk–fibre composites can be produced to order. Special silk–fibre composites might be used in microelectronics and fibre optics or as 'smart' structural fabrics with anti-static properties. Electrostatic properties may also lead to a first market for the more complex mini-machine silks of the capture thread type, be they of the droplet or woolly kind, and they might find employment in active filters which, after all, is an area where orb webs already excel.

Whether the silk material is to be used as an individual fibre or woven into a textile structure, it is critical that production techniques are developed to generate long lengths of material in sufficient quantity and with mechanical performance that is at least equal to the native spider silks. Traditionally, silkworm silks have been the focus of research to generate silk–fibre materials. Techniques to form fibres from silkworm proteins, such as solvent extrusion, electrospinning and microfluidic approaches, might be appropriate for spider-silk proteins as well. The advantages and limitations of each system might determine their use in specific applications or their commercial exploitation.

Current research in spider silk involves its potential use as an incredibly strong and versatile material. The interest in spider silk is mainly due to a combination of its mechanical properties and the non-polluting way in which it is made. The production of modern manmade super-fibres such as Kevlar involves petrochemical processing which contributes to pollution. Kevlar is also drawn from concentrated sulphuric acid. In contrast, the production of spider silk is completely environmentally friendly. It is made by spiders at ambient temperature and pressure and is drawn from water. In addition, silk is completely biodegradable. If the production of spider silk ever becomes industrially viable, it could replace Kevlar and be used to make a diverse range of items such as:

- bullet-proof clothing;
- wear-resistant lightweight clothing;
- ropes, nets, seat belts, parachutes;
- rust-free panels on motor vehicles or boats;

- biodegradable bottles;
- bandages, surgical thread and
- artificial tendons or ligaments, supports for weak blood vessels.

One potential application of spider silks is to emulate the diverse material functions of this family of proteins as a source of novel biomaterial designs. Insight into the assembly and processing of spider-silk proteins into various material forms has been a longstanding focus and has allowed the broadening of the field of applications for silks in general. Specifically, medical devices and tissue engineering applications are perhaps the most promising areas for the utilization of spider silks. Recent progress with reprocessed or native silkworm silk fibres has been realized, and similar approaches could be used with spider silks when they become available in sufficient quantities. For example, in ligament-tissue engineering, a combination of fibre twisting and braiding of silkworm silk fibres was able to direct stem-cell- and ligament-cell-based reconstruction through the alignment and mechanical strength of the twisted silk structure (Altman *et al.*, 2002, 2003). Recently, spider silk fibres were manually collected from the major ampullate dragline and seeded with human Schwann cells to demonstrate biocompatibility, suggesting a promising strategy for future treatment of peripheral nerve injuries (Allmeling *et al.*, 2006).

Films have been cast using HFIP solutions that contain recombinant forms of the dragline spider-silk proteins ADF-3 and ADF-4. The transparent films, ranging in thickness from 0.5 to 1.5 mm, could be made water-insoluble through the use of potassium phosphate or methanol, which converted the protein's secondary structure from ahelix to β-sheet. When this treatment was applied to the recombinant form of the ADF-4 protein, a level of chemical stability in certain denaturants was achieved that rivalled the stability of native dragline silk. Cloning strategies and the ability to modify the film surfaces for attachment of functional molecules make these materials promising for applications such as wound dressings and enzyme immobilization scaffolds (Huemmerich *et al.*, 2006; Junghans *et al.*, 2006).

Hydrogels are utilized in the field of tissue engineering to form porous but stable tissue scaffolds. By adding methanol to recombinant forms of the dragline silk protein ADF-4, the silk has been shown to self-assemble into nanofibres with diameters of 3 nm and lengths less than 1 mm. Over the course of a few days, these nanofibres transformed into a hydrogel fibre network. These hydrogels exhibited a non-linear viscoelastic material response with low stiffness and strength. Cross-linking was induced by visible light after applying ammonium peroxodisulfate and tris (2,20-bipyridyl) dichlororuthenium (II) to the surface of the hydrogel. Cross-linked hydrogels demonstrated fairly linear material response with much greater modulus and strength response.

Given the superior mechanical response and the stability of the hydrogel over weeks, these materials are suitable for tissue engineering applications (Rammensee *et al.,* 2006). Sponge-like, porous three-dimensional structures are also important in tissue engineering. Such structures act as a scaffold to support cells, allow transport of nutrients and metabolic wastes and promote tissue development (Nazarov *et al.*, 2004; Kim *et al.*, 2005).

While development of silkworm silk-based sponges has progressed for such applications, the use of spider silk has lagged behind. In one application of spider silk, spider cocoon silk was used to create porous scaffolds for tissue engineered cartilage. Through a process of solubilizing of the cocoon silk in lithium bromide (LiBr), mixing with salts and methanol treatment, porous scaffolds were generated that sustained chondrocyte cell growth, which could lead to articular cartilage development (Gellynck *et al.*, 2005).

Microcapsules of the recombinant form of ADF-4 spider silks were developed by controlling silk self-assembly at an oil–liquid interface. The resulting β-sheet-rich thin polymer shells had high mechanical stability, with wall thicknesses on the order of 50 nm and diameters between 1 and 30 mm. In addition, the properties of the microcapsules, including constrained degradation response to tissue-specific enzymes, were easily controlled. These materials, therefore, offer promise for a range of applications, from drug delivery to microreactor design (Gellynck *et al.*, 2005).

7.8 References

Allmeling, C., Jokuszies, A., Reimers, K., Kall, S. and Vogt, P.M. (2006), Use of spider silk fibres as an innovative material in a biocompatible artificial nerve conduit, *J. Cell. Mol. Med.*, **10**, 770–777.

Altman, G.H., Diaz, F., Jakuba, C., Calabro, T., Horan, R.L., Chen, J., Lu, H., Richmond, J. and Kaplan, D.L. (2003), Silk-based biomaterials, *Biomaterials*, **24**, 401–416.

Altman, G.H., Horana, R.L., Lua, H.H., Moreaua, J., Martinb, I., Richmondc, J.C. and Kaplana, D.L. (2002), Silk matrix for tissue engineered anterior cruciate ligaments, *Biomaterials*, **23**, 4131–4141.

Arcidiacono, S., Mello, C.M., Butler, M., Welsh, E., Soares, J.W., Allen, A., Ziegler, D., Laue, T. and Chase, S. (2002), Aqueous processing and fiber spinning of recombinant spider silks, *Macromolecules*, **35**, 1262–1266.

Ayoub, N.A., Garb, J.E., Tinghitella, R.M., Collin, M.A. and Hayashi, C.Y. (2007), Blueprint for a high-performance biomaterial: full-length spider dragline silk genes, *PLoS ONE*, **2**, e514.

Beckwitt, R. and Arcidiacono, S. (1994), Sequence conservation in the Cterminal region of spider silk proteins (Spidroin) from *Nephila clavipes* (Tetragnathidae) and *Araneus bicentenarius* (Araneidae), *J. Biol. Chem.*, **269**(9), 6661–6663.

Bini, E., Knight, D.P. and Kaplan, D.L. (2004), Mapping domain structures in silks from insects and spiders related to protein assembly, *J. Mol. Biol.* **335**, 27–40.

Blackledge, T.A., Scharff, N., Coddington, J.A., Szuts, T., Wenzel, J.W., Hayashi, C.Y. and Agnarsson, I. (2009), Reconstructing web evolution and spider diversification in the molecular era, *Proc. Natl. Acad. Sci. USA*, **106**, 5229–5234.

Bunning, T.J., Jiang, H., Adams, W.W., Crane, R.L., Farmer, B. and Kaplan, D. (1994), Applications of silk. In *Silk Polymers—Materials Science and Biotechnology*, ed. Kaplan, D., Adams, W.W., Farmer, B. and Viney, C., American Chemical Society: Washington, DC, **544**, 353–358.

Casem, M.L., Tran, L.P. and Moore, A.M. (2002), Ultrastructure of the major ampullate gland of the black widow spider, *Latrodectus Hesperus*, *Tissue Cell*, **34**(6), 427–436.

Casem, M.L., Turner, D. and Houchin, K. (1999), Protein and amino acid composition of silks from the cob weaver, *Latrodectus hesperus* (black widow), *Int. J. Biol. Macromol.*, **24**(2–3), 103–108.

Champion de Crespigny, F.E., Herberstein, M.E. and Elgar, M.A. (2001), Food caching in orb-web spiders (*Araneae: Araneoidea*), *Naturwissenschaften*, **88**(1), 42–45.

Craig, C.L., Riekel, C., Herberstein, M.E., Weber, R.S., Kaplan, D. and Pierce, N.E. (2000), Evidence for diet effects on the composition of silk proteins produced by spiders, *Mol. Biol. Evol.*, **17**(12), 1904–1913.

Cunniff, P.M., Fossey, S.A. and Auerbach, M.A. (1994), *Poly. Adv. Technol.*, **5**, 401.

Cunniff, P.M., Fossey, S.A., Auerbach, M.A. and Song, J.W. (1994), Silk polymers: materials science and biotechnology. In *American Chemical Society Symposium Series*, **544**, 34, http://pubs.acs.org/books/publish.shtml.

Denny, M.W. (1980), Silks-their properties and functions, In *Mechanical Properties of Biological Materials*, ed. Vincent, J.F.V. and Currey, J.D., Cambridge University Press, New York, 247.

Donald, A.M. and Windle, A.H. (1992), *Liquid Crystalline Polymers*, **1**, Cambridge University Press, Cambridge.

Edmonds, D.T. and Vollrath, F. (1992), *Proc. R. Soc. London, B*, **248**, 145.

Exler, J.H., Hümmerich, D. and Scheibel, T. (2007), The amphiphilic properties of spider silks are important for spinning, *Angew. Chem. Int. Ed. Engl.*, **46**, 3559–3562.

Freddi, G. and Tsukada, M. (1996), *Polymeric Materials Encyclopedia*, 7734, CRC Press, Boca Raton, FL.

Garb, J.E., DiMauro, T., Lewis, R.V. and Hayashi, C.Y. (2007), Expansion and intragenic homogenization of spider silk genes since the triassic: evidence from mygalomorphae (tarantulas and their kin) spidroins, *Mol. Biol. Evol.*, **24**, 2454–2464.

Garrido, M.A., Elices, M., Viney, C. and Pérez-Rigueiro, J. (2002), *Polymer*, **43**, 1537.

Gellynck, K. (2005). Chondrocyte growth in porous spider silk 3D scaffolds, *Eur. Cell. Mater.* **10**(2), 45.

Gerritsen, V.B. (2002), The tiptoe of an airbus, *Prot. Spotlight Swiss Prot.*, **24**, 1–2.

Gosline, J.M., DeMont, M.E. and Denny, M.W. (1986), The structure and properties of spider silk, *Endeavour*, **10**, 37.

Gosline, J.M., Denny, M.W. and DeMont, M.E. (1984), Spider silk as rubber, *Nature*, **309**, 551–552.

Gosline, J.M., Guerette, P.A., Ortlepp, C.S. and Savage, K.N. (1999), The mechanical design of spider silks: from fibroin sequence to mechanical function, *J. Exp. Biol.* **202**, 3295–3303.

Gould, P. (2002), Exploiting spider's silk, *Materials Today*, **5**(12), 42–47.

Guerette, P. The Mechanical Properties of Spider Silk are determined by the genetic regulation of fibroin proteins and chemical and physical processing during spinning. http/www.zoology.ubc.ca/labs/biomaterials/ab-paul.html, accessed July 2005.

Hayashi, C.Y., Shipley, N.H. and Lewis, R.V. (1999), Hypotheses that correlate the sequence, structure, and mechanical properties of spider silk proteins, *Int. J. Biol. Macromol.*, **24**, 271–275.

Heim, M., Romer, L. and Scheibel, T. (2010), Hierarchical structures made of proteins: the complex architecture of spider webs and their constituent silk proteins, *Chem. Soc. Rev.*, **39**, 156–164.

Hinman, M.B. and Lewis, R.V. (1992), Isolation of a clone encoding a second dragline silk fibroin. *Nephila clavipes* dragline silk is a two-protein fibre, *J. Biol. Chem.*, **267**(27), 19320–19324.

Horan, R.L., Antle, K., Collette, A.L., Wang, Y., Huang, J., Moreau, J.E., Volloch, V., Kaplan, D.L. and Altman, G.H. (2005), *In vitro* degradation of silk fibroin, *Biomaterials*, **26**, 3385–3393.

Huemmerich, D., Scheibel, T., Vollrath, F., Cohen, S., Gat, U. and Ittah, S. (2004), Novel assembly properties of recombinant spider dragline silk proteins, *Curr. Biol.*, **14**, 2070–2074.

Huemmerich, D., Slotta, U. and Scheibel, T. (2006), Processing and modification of films made from recombinant spider silk proteins, *Appl. Phys. A*, **82**, 219–222.

Jensen, M. Gene cloned for stretchiest spider silk, http://www.sciencenews.org/sn_arc98/2_21_98/fob2.htm.

Jin, H.J. and Kaplan, D.L. (2003), Mechanism of silk processing in insects and spiders, *Nature*, **424**, 1057–1061.

Junghans, F., Morawietz, M., Conrad, U., Scheibel, T., A. Heilmann and Spohn, U. (2006), Preparation and mechanical properties of layers made of recombinant spider silk proteins and silk from silk worm, *Appl. Phys. A*, **82**, 253–260.

Kaplan, D., Adams, W., Farmer, B. and Viney, C. (eds) (1994), *Silk Polymers: Materials Science and Biotechnology, American Chemical Society*. American Chemical Society, Washington DC.

Kaplan, D.L., Mello, C.M., Arcidiacono, S., Fossey, S., Senecal, K. and Muller, W. (1998), *Protein Based Materials*, Birkhauser, Boston.

Kim, U.J., Park, J., Kim, H.J., Wada, M. and Kaplan, D.L. (2005), Three-dimensional aqueous-derived biomaterial scaffolds from silk fibroin, *Biomaterials*, **26**, 2775–2785.

Kluge, J.A., Rabotyagova, O., Leisk, G.G. and Kaplan, D. L. (1997), Spider silks and their applications, *Trends in Biotechnology*, **26**(5), 244–251.

Knight, D.P. and Vollrath, F. (1999), *Proc. R. Soc. Lond. B*, **266**, 519.

Ko, F.K., Kawabata, S., Inoue, M., Niwa, M., Fossey, S. and Song, J.W. (2005), www.web.mit.edu/course/3/3.064/www/slides/Ko_spider_silk.pdf.

Kohler, T. and Vollrath, F. (2005), *J. Exp. Zoo*, **271**, 1.

Lewis, R.V. (1992), Spider silk: the unraveling of a mystery, *Acc. Chem. Res.*, **25**(9), 392–398.

Lin, L.H., Edmonds, D.T. and Vollrath, F. (1995), Structural engineering of an orb-spider's web, *Nature*, **373**, 146–148.

Lombardi, S. and Kaplan, D.L. (1990), The amino acid composition of major ampullate gland silk (dragline) of *Nephila clavipes* (*Araneae, Tetragnathidae*), *J. Arachnol.*, **18**, 297–306.

Madsen, B., Shao, Z. and Vollrath, F. (1999), *Int. J. Biol. Macromol.*, **24**, 301.
Madsen, B. and Vollrath, F. (2000), *Naturwissenschaften*, **87**, 148.
Maynes, E., Mann, S. and Vollrath, F. (1998), *Adv. Mater.*, **10**, 801.
Mukhopadhyay, S. and Sakthivel, J.C. (2005), *J. Ind. Text.* **35**, 91.
Nazarov, R., Jin, H.J. and Kaplan, D.L. (2004), Porous 3-D scaffolds from regenerated silk fibroin, *Biomacromolecules*, **5**, 718–726.
Northolt, M.G. and Sikkema, D.J. (1991), *Adv. Polym. Sci.*, **98**, 115.
O'Brien, J.P., Fahnestock, S.R., Termonia, Y. and Gardner, K.C.H. (1998), Nylons from nature: Synthetic analogs to spider silk, *Adv. Mater.*, **10**, 1185–1195.
Rammensee, S., Huemmerich, D., Hermanson, K.D., Scheibel, T. and Bausch, A.R. (2006), Rheological characterization of hydrogels formed by recombinantly produced spider silk, *Appl. Phys. A*, **82**, 261–264.
Scheibel, T. (2004), Spider silks: recombinant synthesis, assembly, spinning, and engineering of synthetic proteins, *Microb. Cell Fact.*, **3**, 14.
Schulz, S. and Toft, S. (1993), Branched long chain alkyl methyl ethers: a new class of lipids from spider silk, *Tetrahedraon*, **49**(31), 6805–6820.
Shao, Z. and Vollrath, F. (1999), The effect of solvents on the contraction and mechanical properties of spider silk, *Polymer*, **40**, 1799.
Shao, Z. and Vollrath, F. (2002), Surprising strength of silkworm silk, *Nature* **418**(6899),741.
Shao, Z., Wen Hu, X., Frische, S. and Vollrath, F. (1999), Heterogeneous morphology in spider silk and its function for mechanical properties, *Polymers*, **40**, 4709.
Sponner, A., Unger, E., Grosse, F. and Weisshart, K. (2005), Differential polymerization of the two main protein components of dragline silk during fibre spinning, *Nat. Mater.*, **4**(10), 772–775.
Sutherland, T.D., Young, J.H., Weisman, S., Hayashi, C.Y. and Merritt, D.J. (2010), Insect silk: one name, many materials, *Ann. Rev. Entomol.*, **55**, 171–188.
Tirrell, J.G., Fournier, M.J., Mason, T.L. and Tirrell, D.A. (1994), *Biomol. Mater. Chem. Eng. News*, **72**, 40.
Tso, I.M., Wu, H.C. and Hwang, I.R. (2005), Giant wood spider *Nephila pilipes* alters silk protein in response to prey variation, *J. Exp. Biol.*, **208**(6), 1053–1061.
Van Beek, J.D., Hess, S., Vollrath, F. and Meier, B.H. (2002), The molecular structure of spider dragline silk: folding and orientation of the protein backbone, *Proc. Natl. Acad. Sci. U. S. A.*, **99**, 10266–10271.
Vollrath, F. (1999), Biology of spider silk, *Int. J. Biol. Macromol.*, **24**, 81.
Vollrath, F., Hum, W. and Knight, D.P. (1998), Silk production in a spider involves acid bath treatment, *Proc. R. Soc. B*, **263**, 817–820.
Vollrath, F. and Knight, D.P. (2002), Liquid crystalline spinning of spider silk, *Nature*, **410**, 541–548.
Work, R.W. (1977a), Mechanisms of major ampullate silk fibres of orb-web spinning spiders, *Trans. Am. Microsc. Soc.*, **100**, 1–20.
Work, R.W. (1977b) *J. Exp. Biol* ., **118**, 379–404.
Xu, M. and Lewis, R.V. (1990), Structure of a protein superfiber: spider dragline silk, *Proc. Natl. Acad. Sci. U.S.A.*, **87**, 7120–7124.
Yoshimizu, H. and Asakura, T. (1990), *J. Appl. Polymer Sci.*, **40**, 127.
Zhang, F., Zhao, Y., Chen, X., Xu, A.Y., Huang, J.T. and Lu, C.D. (1999), Fluorescent transgenic silkworm, *Acta Biochem. Biophys.*, **23**, 119.

Index

CPSIA information can be obtained at www.ICGtesting.com
Printed in the USA
BVOW02*0353030214

343683BV00005B/113/P